When
HEAVEN *and* EARTH
EMBRACE

Mary Tinney offers essential ecological principles born of years of passionate commitment to caring for creation. Her work is both a spirituality and a practice grounded in the faith and hope of Christian theological reflection. This is a work for all those who are lovers of God and the gifts of God's creation.

Terry A. Veling, Australian Catholic University.

Mary Tinney has long been deeply involved in ecological practice and spirituality. In this new book she brings her experience in ecological spirituality into a synthesis with recent ecological theology, and offers us all a vision that can support ecological action. This is a practical and useful book that will be valuable for many searchers in the midst of the crisis of our planetary community.

Denis Edwards (Late), Professorial Fellow in Theology. He was a priest of the Archdiocese of Adelaide and his most recent research was in the areas of Christology, Pneumatology, Trinitarian theology, the dialogue between science and theology, and ecological theology.

The title of this book like its content is active and engaging, inviting readers into the embrace of Heaven and Earth and catching them up into spiritual engagement in an emerging universe. Its author, Mary Tinney, does this through a process she describes as "bringing the craft of practical theology to bear on ecospirituality" thereby providing a way for others to further what she has begun in an age of ecological conversion.

Elaine Wainwright, Professor Emerita of the University of Auckland. She is currently writing the Matthew volume in the Wisdom Commentary series, a feminist commentary on each biblical book, being published by Collegeville Liturgical Press and has recently published *Habitat, Human and Holy: An Eco-Rhetorical Reading of the Gospel of Matthew* in the Earth Bible Commentary Series.

When
HEAVEN *and* EARTH
EMBRACE

How do we Engage Spiritually in an Emerging Universe?

MARY TINNEY

Published by Acorn Press
An imprint of Bible Society Australia
ACN 148 058 306 | Charity licence 19 000 528
GPO Box 4161
Sydney NSW 2001
Australia
www.acornpress.net.au | www.biblesociety.org.au

ISBN 978-0-647-53352-9

First published by Morning Star Publishing in 2019, ISBN 978-0-648-45388-8

 A catalogue record for this work is available from the National Library of Australia

Cover and text design and layout by John Healy

Cover Painting: *The Embrace*, used by permission of the artist, Evelyn Dunphy, www.evelyndunphy.com

Contents

List of Figures

Preface

This book is based on my doctoral thesis of the same name, which was completed in 2018. I believe that this work makes a significant contribution to the fields of theology and spirituality. It has the potential to enhance the praxis of those from the Christian faith tradition who have a passionate concern about the wellbeing of Earth and cosmos.

The following pages offer a theologically informed approach to ecospirituality, supplemented by guides for reflection and conversation. The book is part of the unfolding journey of Earth Link, a community group that came into existence in 2000 and continues to the present day, although on a reduced scale. It outlines some of that journey, with particular emphasis on the group's distinctive approach to ecospirituality. The Earth Link community has been at the heart of all developments, and its core group, whose membership has changed a little over time, has provided a community of concern for Earth and cosmos, and demonstrated fidelity to exploring the faith and practical implications of such concern.

I wish to acknowledge the support of the Sisters of Mercy of the Brisbane Congregation, and of the Institute of the Sisters of Mercy of Australia and Papua New Guinea, who enabled my dreams and those of the Earth Link community to come into effect. Without their affirmation and resourcing there would be no story to be told. I wish also to acknowledge the contribution of Denis Edwards, who died suddenly a week after endorsing this work. His contribution to the field of ecotheology will continue to be valued globally.

Writing such as this is of necessity a solitary pursuit, but there have been a few unexpected companions on the journey: the Pale-headed Rosella who made a one-off appearance in my garden as I discerned whether to begin the project; the Grey Shrike thrush, whose song marks its arrival around October each year; and the Caper White butterflies who timed their once-in-about-ten year migration to coincide with my preparing the final stage of the thesis. It was truly heartening to take my coffee breaks in their company.

All of these keep reminding me of my identity within the community of creation.

CHAPTER 1

Setting the scene

Earth is in crisis. We need a Christian spiritual praxis[1] that is relevant and possible, one that motivates and sustains us for action. At the heart of such a praxis is the way we envisage the meeting of Heaven and Earth.

For too long the human religious trajectory has been one of escape from our earthly reality to the other-worldly heavenly realms. While Heaven and Earth do meet in that vision, they are pitted against each other, and the emphasis is on the inferiority of the earthly to the heavenly and on the inferiority of nature to the human. There is every indication that this dualistic and hierarchical world view,[2] which underpins much of Western culture, has led us to a disastrous state of affairs. We are in the midst of a global crisis, in which climate change and environmental degradation need to concern us all.[3] The data is to hand that human influence is contributing seriously to these problems.[4] Our vision of the relationship between Heaven and Earth needs to change.

Scientific findings of the twentieth century onwards tell us of the evolutionary or emerging nature of the universe, and of the interdependence and interconnectedness of all reality. For some, the world view of scientific naturalism provides a vision that can take us into the future, but I find this world view inadequate because it depends only on what we can see in a material sense. Neither can I accept a

---

1 Praxis differs from practice by having a considered theoretical underpinning.
2 The Oxford Dictionary defines world view as 'a particular philosophy of life or conception of the world', http://www.oxforddictionaries.com/definition/english/world-view
3 Pope Francis, *Laudato Si.* In Chapter 1 of the encyclical Pope Francis identifies pollution, climate change, water, loss of biodiversity, decline in the quality of human life, the breakdown of society and global inequality as issues that impact negatively on Earth, our common home.
4 Intergovernmental Panel on Climate Change (IPCC), 'Climate Change 2014', p. 2. The first finding is that 'human influence on the climate system is clear, and recent anthropogenic emissions of greenhouse gases are the highest in history. Recent climate changes have had widespread impacts on human and natural systems.'

world view that denigrates Earth in the pursuit of a heavenly future, or one in which Heaven and Earth completely overlap without allowing for differences. This book holds up a vision that links the material and the spiritual in a way that values both Heaven and Earth, recognising their embrace as the lynchpin for engaging spiritually in our emerging universe so that our spiritual and ethical practices move us towards redressing environmental degradation and human poverty.

In this book I argue that it is possible for us to engage spiritually in an emerging universe if we have a vision of the embrace of Heaven and Earth[5] that is informed by contemporary science, if we underpin this vision with an ecotheology that recognises Heaven and Earth as interconnected while respecting their differences, and if we cultivate an open ecospiritual praxis in which we are attentive to, and aware of, divine presence in all that is. This is not the dominant paradigm. To adopt this new paradigm we will need to re-vision many assumptions.

Heaven and Earth do not just meet at every moment – they embrace, always interdependent, together on a journey that is constantly unfolding. My image for this embrace is the Celtic triquetra,[6] which heads this and most of the subsequent chapters. There is a connection, an intimacy, an embrace, which allows for the distinctiveness of the parts. Such a world view allows each part to 'stand in its own difference, but encompassed by a wider whole that affects their interrelatedness.'[7] No part is complete except as part of the whole, and the whole 'transcends yet includes'[8] the parts. We engage spiritually in an emerging universe as a part within an interconnected whole. We are part of the mutual engagement of the parts with one another and with the whole.

The use of the Celtic triquetra also underlines that the parts are interconnected in their distinctiveness, and not in any sense of hierarchical

---

5 References to Heaven and Earth, Planet Earth and the Earth community will be capitalised in this book.

6 There are many versions of the Celtic knot, but the one I am using has a circle going through the three parts of the triquetra or triangle, underlining the interconnectedness and unity of the three parts. The image is available in the public domain.

7 Johnson, *Ask the Beasts*, p. 269.

8 Phipps, *Evolutionaries,* p. 191. Phipps acknowledges Hegel's recognition of this pattern of 'transcending and including' as a universal principle of evolutionary emergence.

value. At the same time, there is no hint that the parts are equal. While the title of the book contains only two elements, Heaven and Earth, we will at times consider humans as the third element of the triquetra when looking at Earth as the whole Earth or planetary community, and at other times separately as a species of greater complexity, consciousness and dignity. As a symbol of the Trinity, the use of the Celtic triquetra also sets the tone for the theological reflections.

In the context of this book, Heaven[9] is the locus or place of the Divine.[10] Heaven is not understood as an other-worldly realm[11], but rather as a relational reality that is mediated by the physical, planetary and cosmic realms yet opens up towards fullness of life in the Divine. It is the encounter with the Divine that is significant. The place and nature of the encounter will feature prominently in this book. My approach to the Divine derives from contemporary understandings of God, whose transcendence and immanence are intimately connected. At different times in the course of this book, I refer to God in various instances as God, Ultimate Mystery, the Divine and the Sacred. This is part of a sensitivity to those whom I have encountered who are experiencing what Ilia Delio calls a 'new atheism,'[12] a necessary phase of letting go of one God and making way for another. People can be averse to naming God as God because of negative connotations they are seeking to revise.

When I refer to 'Earth'[13], I do not mean the soil but the planet, including the whole Earth community, to which human society belongs[14], and all

---

9 Heaven is defined in the Oxford Dictionary as 'a place regarded in various religions as the abode of God (or the gods) and the angels, and of the good after death, often traditionally depicted as being above the sky,' http://www.oxforddictionaries.com/definition/english/heaven. This is not the sense in which the term is used here.

10 When I am using the Divine as a naming of God, I will use upper case.

11 Heaven is also referred to as the doorway to the cosmos, the galaxies, planets and stars, the context in which we live out the drama of life. Heaven will not be used in this sense.

12 Delio, *Unbearable Wholeness.* Chapter 4 is about birthing a new God.

13 In the Oxford Dictionary, 'Earth' is understood as the planet on which we live, and 'earth' is the soil, or substance of the land surface, http://www.oxforddictionaries.com/definition/english/earth. I will use the term to mean the planet, including the Earth community.

14 Rasmussen, *Earth Community,* p. 9. Rasmussen coined the term 'Earth community'" to speak of the '*internal*, not external, linkage of society and environment, and of nature, history and culture.'

parts of which are caught up in processes of evolutionary emergence.[15] Planet Earth is also part of the unfolding story of the universe and thus has a cosmic[16] dimension. This perspective influences the formation of an anthropocosmic[17] world view, which I explore for its theological meaning using the work of Elizabeth Johnson. It is the whole that is in relationship with the Divine, and this is the basis of a theocentric world view.

This book is about how we humans engage spiritually within an emerging universe. While the material is applicable generally, the focus is on one community group and the way it responds spiritually to the current environmental crisis. This book is grounded in the principles and practice of the Earth Link community, which since 2000 has been educating, reflecting, resourcing and acting towards its vision of a 'world where there is respect, reverence and care for the whole Earth community.' Members of the Earth Link community recognise that 'spirituality comes from the transformative experience of deep bonding with Earth,'[18] and their experience has led them to believe that this is an experience of the Sacred. Earth Link is an initiative of the Sisters of Mercy within the Catholic Church as part of their commitment to 'extravagant hospitality, justice and compassion in the Earth community, shattered by displacement.'[19]

The journey described in the book takes the Earth Link community to a new place as it engages spiritually in our emerging universe. Spirituality is not necessarily aligned with a religious tradition, as we read in the words of Bron Taylor: 'Spirituality can be understood as a quest to deepen, renew, or tap into the most profound insights of traditional religions, as well as a word that consecrates

---

15 Edwards, *Partaking of God*, p. 75. The term 'evolutionary emergence' provides a concise way of referring to the dynamics of the evolutionary process.

16 In the Oxford Dictionary , 'cosmos' is understood as the 'universe seen as a well-ordered whole', https://en.oxforddictionaries.com/definition/cosmos.

17 Grim & Tucker, *Ecology and Religion*, p. 56. The authors explain that 'anthropocosmic views of the human as emerging from out of the processes of nature provide new orientations for mutually enhancing human-earth relations of participation rather than domination.'

18 Costigan, Rose & Tinney, *Introduction to Ecospirituality*. This publication was a series of essays on the principles of Earth Link, developed in 2007 by Dr Philip Costigan, Dr Patricia Rose and Sr Mary Tinney.

19 Statement of the Sisters of Mercy of Australia and Papua New Guinea, Chapter, 2011.

otherwise secular endeavours such as psychotherapy, political and environmental activism, and one's lifestyle and vocational choices.'[20]

However, Earth Link engages directly with the Christian tradition and its contemporary approaches to ecospirituality and ecotheology, while acknowledging the contribution of secular approaches. Thomas Berry, as one of the key influences on the formation of Earth Link, continues to feature prominently in this consideration of ecospirituality.

Ecospirituality needs to be grounded in well-considered ecotheology. The work of Elizabeth Johnson is very pertinent in this respect. Ecospirituality also needs to be relevant to the reality we live in, namely, that of an evolving, emerging universe. Ecospirituality and ecotheology need to be cognisant of developments in contemporary science and philosophy, and the implications that flow from them. Ecotheology is important for the enhanced ecospiritual principles and practice presented in this book, which I have developed for the Earth Link community.

Outline

In Chapter 2, I begin with an introduction to Earth Link, its origins and what it stands for. The Earth Link community was the springboard for the thesis that forms the basis of much of this book. In Chapter 3, I first indicate how practical theology is an appropriate methodology for establishing connections between life and theology, then go on to summarise the dialogues with ecospirituality and ecotheology, and the affirmations and challenges, that my research opened up. In the main part of the book, in Chapters 4 to 8, I expound an enhanced version of the vision, mission and principles for Earth Link. I supplement this exposition with suggestions for reflection and conversation that can help to integrate the material into the praxis of individuals and groups who share a passionate concern for the wellbeing of Earth and cosmos, based in a vision of the embrace of Heaven and Earth.

---

20 Taylor, *Dark Green Religion*, p. 3.

CHAPTER 2

Introducing Earth Link

Earth Link has a vision of a world
where there is respect, reverence and care
for the whole Earth community.[1]

While the subject matter of this book is relevant to anyone who takes seriously their commitment to the wellbeing of planet Earth and the cosmos, I begin in this chapter with the specific group at the centre of this research, namely, Earth Link. What follows is a description of the genesis and development of the Earth Link project, before a consideration, in some depth, of what Earth Link stands for and how it gives expression to its vision and mission.

Genesis and development of Earth Link

As I was the instigator of Earth Link, its genesis is very much intertwined with my story. My background is in education, especially in religious education, and in leadership of the Sisters of Mercy, a congregation of women within the Catholic Church. My attempts to come to terms with the diminution of religious orders had awoken in me an awareness of the paradigm shifts that were and are happening around us. I was conversant with shifts in the understanding of the building blocks of matter, the atoms. Whereas atoms had been previously understood to be composed of particles rotating in fixed orbits around a nucleus, they were now known to be much more dynamic in their movement, with the particles moving in and out from potential to reality. Findings from physics were driving a shift from a static, mechanistic world view to a more dynamic, ecological world view in which the interconnectedness of all of reality was recognised and the only constant was change. I

1 Vision Statement, Earth Link www.earth-link.org.au. The image at the beginning of this chapter is Earth Link's logo.

had explored the implications for organisational life of these changing understandings and the world views built on them. Nothing prepared me for the shock of discovering that the human-centred or anthropocentric world view dominant in Western society could, in fact, be a major contributing factor to the deteriorating state of the environment. What was more shocking was my growing awareness that the Judaeo-Christian tradition, which reinforced the notion of human superiority over the rest of created reality, was seen by some as being a major contributor to the ecological crisis.[2]

As part of my response to this, I have been living since the turn of the century out of a deepening consciousness of my place in the universe, which, according to current data, is about 13.75 billion years old. When I first recognised my place in this deep history, I felt as if I had been hit by an asteroid. That encounter was hardly an event of global proportions, but I did feel as if my centre of gravity had shifted. My human-centred or anthropocentric world view began to give way to a more inclusive, life-centred world view, with the subsequent need to review most of the assumptions that had given me meaning for many years. Some of these experiences needed to be available to others. With the endorsement and financial support of the Sisters of Mercy, I set up Earth Link to offer educational and reflective programmes. After about two years, a sixteen-hectare property, Four Winds, in the district of Ocean View[3], became available. It had accommodation for eight people. As of this point Earth Link had a home.

In the course of the following year, 2002, a group of people with backgrounds in environmental and religious education designed a curriculum for a four-unit program covering cosmology, sense of place (bioregionalism), action for justice for Earth, and immersion experiences in other bioregions. The program was initially offered on site, while workshops derived from the content were offered more widely. The courses were eventually made available in distance mode. We promoted the initiatives on the Earth Link website[4] and distributed

---

2 See White, 'Historical Roots'.

3 Ocean View is approximately one hour's drive from Brisbane, Queensland, Australia.

4 www.earth-link.org.au.

flyers by hand and on community noticeboards. Specific events were advertised in local and regional newspapers. In 2002 I was able to attend the United Nations Climate Summit held in Johannesburg, absorbing its global perspective and observing Australia's embarrassing role in the negotiations.

Apart from developing the program we devoted time to managing the property in an environmentally responsible manner. Its location twelve kilometres from the nearest small township meant that we were self-sufficient in terms of water and responsible for our own waste disposal; we had a greywater system and an onsite septic system. Our attempts to live sustainably led us to establish a permaculture garden, and eventually to install solar power and solar hot water systems, as well as an outdoor composting toilet for use by groups who visited or came for programs. The biggest challenge was the care of the fifteen hectares that had not been cleared. This was, and still is, an important reserve within the local bioregion. Learning how to maintain this land under the 'Land for Wildlife' scheme[5] was a challenge, especially how to manage weeds such as lantana, which seemed to have the perfect growing environment in our dry sclerophyll forest. Fortunately, there was federal funding available for such efforts, and we were successful in our applications for grants over several years.

In 2004, the three-yearly review of the programme indicated a stronger interest in ecospirituality. In 2005, a community of committed persons gathered to read, study and discuss this topic, before commissioning a writing group, who became the Spirituality Team. During this period, thanks to grant funding, we developed resources for use in personal reflection and in workshops. We developed our website and began the production of a monthly online newsletter that included reflections and book reviews. Ritual spaces and walks were developed on the property; these featured in our workshops and were available to guests. The ecospirituality principles developed in this period were published in a booklet entitled *Introduction to Ecospirituality.*[6] This booklet

---

5 Land for Wildlife is a voluntary program auspiced by local governments that encourages and assists landholders to manage wildlife habitat on their properties. See https://www.lfwseq.org.au.

6 Costigan, Rose & Tinney, *Introduction to Ecospirituality*.

has proved to be very popular. It has formed the basis of facilitated workshops and self-directed retreats in several places. Resources to accompany the booklet were/are available on request. This program was even integrated into a university module conducted by Australian Catholic University. An article based on our principles was published in the journal *Social Alternatives.*[7] We also contributed a five-part series to the *Earth Song Journal*[8] between 2007 and 2010. From 2005 to 2010 Earth Link was at its peak. We continued to be subsidised by the Sisters of Mercy, while attempting to elicit grant funding, especially for improvements of the property.

The ecospirituality principles developed by Earth Link were always intended to be generic and appropriate for anyone who shared our vision, regardless of their faith tradition. We had by now come to the attention of the Catholic 'thought police', who were particularly active in Brisbane and surrounds between 2004 and 2011. We were reported to the Archbishop in both 2008 and 2011, who in his correspondence with the then Congregation Leader of the Brisbane Sisters of Mercy expressed his support for what we were doing, affirming both its importance and our openness to all, regardless of their faith tradition.[9] In 2004 and 2006, we received adverse publicity in the conservative magazine, *AD 2000*, which comes out of New Zealand. I even received a letter from someone in New York accusing us of worshipping creation rather than the Creator. She suggested that we should be called the Sisters of Recycling. I simply wished that we were as well known by mainstream religious people in the local area as by such detractors. We were always conscious that we were a publicly Catholic Christian ministry of the Sisters of Mercy that was open to all, in the interests of the wellbeing of the whole Earth community, as a response to widespread environmental degradation and human poverty. This was and is a responsibility that we take seriously.

---

7 Social Alternatives is an independent, quarterly, refereed journal that aims to promote public debate, commentary and dialogue about contemporary social. political, economic and environmental issues. See http://socialalternatives.com/
8 The *Earth Song Journal* was produced twice yearly by the Earth Song Project, based in Melbourne. This project has now finished.
9 This correspondence is held in the archives of the Sisters of Mercy, Brisbane Congregation.

It was always intended that we would initiate a dialogue with others in the Christian and multifaith traditions about ecospirituality and its theological underpinnings. I prepared a booklet called *Ecology and Christian Faith*, which highlighted the problems and potential of Christian theology. Workshops on this material were offered initially at Four Winds. Our attempts to find partners who would reflect with us on the connection between ecology and other faith traditions were not very successful at that time, even though we had some minimal involvement with a multifaith centre at a university. We were approved to offer a workshop at the Parliament of the World's Religions held in Melbourne in 2009, but generally the time had not quite arrived for the kind of Brisbane-based multifaith dialogue we had envisaged.

By 2009, it was time for a transition for Earth Link. A report was commissioned that led to the appointment of a part-time property manager and a part-time director who had the potential to move into the program area. Unfortunately, the changeover to new leadership did not work, and it was deemed timely to sell the property. Earth Link moved completely offsite at the end of 2012, at which time I downsized to the current strategies of resourcing, reflecting and acting. A core community continues to meet monthly for education and reflection. The main activities for carrying forward the vision and the mission are the e-newsletter, which has about 900 recipients who form an online community, and the website, which is updated according to the most recent newsletter. I respond to invitations for presentations and workshops, mostly in the area of ecospirituality, and Earth Link offers some public events, such as a workshop on the encyclical of Pope Francis, and a discussion on the work of Naomi Klein. The demand indicates the potential of ecospirituality as a vehicle of transformation. Because of my work with Earth Link, I have also been in a position to act as a resource person for the Institute of the Sisters of Mercy of Australia and Papua New Guinea, especially in the launch of *An Integrated Policy on Sustainable Living*. This activity is outside the scope of this narrative but is opening up new avenues for involvement.

A number of developments over the last few years have changed the level of interest in ecological matters for the better. Since 2011, workshops introducing ecospirituality have been offered by the Australian Earth Laws Alliance (AELA), and these have been very well attended by a wide range of concerned people. The formation in 2011 of the Queensland Churches Environment Network (QCEN) as a commission of Queensland Churches Together provided a Christian ecumenical network that is proving very supportive and effective in promoting a Christian response to concerns for the whole Earth community. The release by Pope Francis of the encyclical *Laudato Si, On Care for our Common Home* in July 2015 brought remarkable energy to the conversation within and across faith traditions, and Earth Link conducted several events around the content of, and responses to, the encyclical.

It would seem that Earth Link's pioneering efforts since 2000 have situated us well for this moment in time, and into the future. I will turn now to consider the nature of the commitment of Earth Link, and some of the assumptions out of which Earth Link operates.

What does Earth Link stand for?

Earth Link has a basic understanding that fostering the embrace of Heaven and Earth, manifested in the relationships between God, Earth and her inhabitants, is important for the wellbeing of Earth and cosmos. This is articulated in its vision, mission and principles.

The website is the public face of Earth Link, and its home page manifests clearly that 'Earth Link is a community which envisions a world where there is respect, reverence and care for the whole Earth community. We believe that the heart of this lies in deepening our bond with Earth.'[10]

Earth Link's mission to facilitate this process is introduced in the preamble to *Introduction to Ecospirituality*:

> In our times, there are ongoing shifts in our knowledge about the unfolding story of the Universe, the nature of matter, the interconnectedness of the web of life, the importance of the ancient Earth traditions and the list goes on. Many people are seeking to

---

10 www.earth-link.org.au

make meaning of these changes to deepen their own connections with nature and the cosmos, to explore the implications for their own religious and spiritual traditions, and to name their experience of the Sacred. As we enter into this journey, we are articulating the spirituality of Earth Link.

We understand that the Earth Link spirituality comes from the transformative experience of deep bonding with Earth. Reflection has led us to believe that this is an experience of the Sacred, and that Earth and cosmos constitute for us a primary revelation of Ultimate Mystery.

We go further to articulate five principles of ecospirituality:

1. Listen to the wisdom of Earth with an open, attentive and receptive attitude.
2. Deepen your relationship with cosmos/Earth, beginning with your own particular place.
3. Acknowledge the Sacred in the interdependent web of life.
4. Honour the Sacred in the web of life through rituals and holistic living.
5. Live in right relationships within the interdependent web of life.[11]

The themes that stand out in the Earth Link principles are listening, place, interdependence, acknowledging and honouring the Sacred in the web of life, and living in right relationships.

In the abstract to an article by the then Spirituality Team of Earth Link, Philip Costigan wrote about the role of spirituality in an ecocentric culture and the contribution made by Earth Link as one response to an environment in crisis:

A spirituality which envisions the Sacred as intimately embodied in the Earth and the cosmos brings into force powerful emotions of reverence for all life and commitment to justice for the Earth. A deep bonding with nature, and recognition that humankind is only one element in the whole interdependent web of life, underpins this type of spirituality. Ecospirituality can offer a solid ideological and

---

11 Costigan, Rose & Tinney, *Introduction to Ecospirituality.*

theological base for the current environmental movement. Because it taps into deep-rooted motivations and commitments it has the power to challenge radically, and change fundamentally, the destructive culture that exploits the earth, and transform it into a culture that is life-enhancing and eco-centric. The ecospirituality put forward by Earth Link encompasses a comprehensive set of experiences, beliefs, rituals and actions, and is one attempt to formulate a spiritual framework for living in ways that are more ecologically sensitive.[12]

The principles of Earth Link highlight the importance of a contemplative attitude to Earth and cosmos, and an open, attentive and listening stance. Earth Link encourages immersion in nature with a receptive attitude. This is based on an awareness that such an encounter is one of subject to subject, rather than subject to object. There is mutuality in the encounter. This 'deep bonding' is key to the approach of Earth Link, and together we have seen that it is the heart of the matter. 'Deep bonding' is the encounter that opens us to an appreciation of the intrinsic worth of nature, both human and other-than-human, and to an experience of the Sacred, however that is understood by the individual.

Ecospirituality, as Earth Link understands it, is a place-based spirituality. This is when my body, the body of Earth and cosmos, and the body of God may meet. The 'deep bonding' mentioned above is bonding with a particular place or places. These are not necessarily wild places, although some such exposure is recommended. They can be where you live or where you have fond memories of your associations with people and place. Many treasured encounters come from childhood, and sometimes remembering those encounters can help or hinder the encounter with one's current place or places. Unless we know a place, we cannot read what it may be saying.

Earth Link is conscious that there are wisdom traditions, religious and otherwise, which give meaning to what we are encountering. These belief and value systems can be normative and provide a community of concern. However, religious systems, like economic and social ones, are often exclusive of environmental appreciation and concerns. Earth Link

---

12 Costigan, Rose & Tinney, 'Role of Spirituality', *Social Alternatives,* vol. 26, no. 3, Third Quarter, 2007.

is conscious that most traditional systems, as mentioned in the section on context, are mired in classical, human-centred or anthropocentric world views. While contemporary Christian theological writers are working more and more within evolution-based, science-aware frameworks, this is not the case in most faith communities. The ecospirituality booklet points to the paradigm shift that is needed if the interconnection of God and the Earth community are to be the basis of a faith response that respects the integrity of the web of life. It points to the agenda of theological re-visioning that is required, without actually developing it.

Earth Link recognises the importance of expressing honour, respect, and esteem for the Sacred, individually and in community, through rituals and holistic living. We set up ritual places and spaces, mostly outdoors. We laid down a Cosmic Walk celebrating the unfolding story of the universe over 13.75 billion years − a foundational story for Earth Link because of its potential for developing a sense of the place of humans in the story of evolution. We marked out safe walks through the property, and prepared booklets that mingled ecological information with the invitation to reflective moments. We prepared books that facilitated reflection on the hours of the day and the seasons of the year. We offered workshops with people like John Seed, an Australian influenced by the Buddhist tradition, who worked with Joanna Macy, an internationally recognised activist, to develop rituals and processes such as the Council of All Beings. In that process/ritual, people are invited to allow some aspect of nature to call them into partnership, and then to represent them at a Council of all Beings, where the humans prepare and don masks suggestive of the being that they represent. This is a very powerful way of getting into the shoes (so to speak) of the other. Earth Link continues, although to a lesser extent, to ritualise around the seasons of the year and the liturgical seasons of the church year. In this way it opens up the time-honoured tradition of *lectio divina*, taking it beyond listening only to the book of scripture to listening to the book of nature as well.

Earth Link sees its strength in the field of attitude change, but its final principle is about living in right relations. We established some criteria for right relations with self, others, Earth, cosmos and the Sacred. These include subject-subject relations with all, an ecocentric world view, relations that are partnerships rather than domination, and engagement in any action 'with' Earth, not 'for' Earth.

As noted in the above section concerning the genesis and development of Earth Link, the project is not without history in the area of ecotheology. As an extension of the ecospirituality principle that invites people to 'acknowledge the Sacred in the interdependent web of life,' I prepared a booklet entitled *Ecology and Christian Faith.* Our work on ecospirituality had identified the need for a theology that integrated Earth, the human and the Sacred. The booklet explores the proposition that 'Christianity is part of the problem of environmental degradation, but also that Christianity contains within it the promise of a better future if it recognises the role that it can play with others who are addressing the environmental crisis.'[13]

The booklet is in two parts: the Earth-human connection and the Earth-human-Sacred connection. In the first section, three problems were noted: first, hierarchical ranking and valuing, and the licence this affords those higher up the ladder to exploit those lower down; secondly, dualistic distinctions between spirit and matter, and between soul and body, with the subsequent devaluing of the latter in each case; and, thirdly, the way in which decision-making in an anthropocentric society favours humans and their economic systems, often at the expense of the environment. Insights from cosmology and evolutionary theory were brought to bear to establish the place of humans within the Earth community. Insights from theology led to exploration of more meaningful Earth-human relations that recognise the distinctiveness of humans among other species, and the limited and finite nature of all reality. We also explored what our ultimate destiny might be like when we return to the dust of Earth.

In the second section, that on Earth-human-Sacred connections, I identified problems associated with the apathy within the Christian community, based on a lack of awareness that our relationship with Earth/universe/cosmos is a dimension of Christian life. Other problems come from an outmode view of the cosmos 'as having been

---

13 Tinney, *Ecology and Christian Faith.* There are no page numbers in this booklet.

launched in a finished form by the hand of the Creator, rather than as a dynamic, unfolding universe in which Earth, humans and the Sacred are integral and connected parts of the process.'[14] Having established the interconnections, I went on to explore panentheism and Catherine Keller's notion of 'radical incarnationalism'[15], which does not blur the distinction between Earth and Divine Mystery but allows for intense and open-ended interaction between them. There is great potential in Christianity if we are willing to 'reformulate our understanding of the Sacred whom we encounter in the midst of our contemporary living situations.'[16] In the rudimentary Trinitarian theology that followed, I considered the God who creates (and destroys), and Jesus who came among us, and God's continuing presence among us as Spirit, drawing us into fullness of Being. This theology explored the first Creation narrative in Genesis 1:1–2:4a as creation from within the unfolding story of the universe rather than as creation from above. It acknowledged other rich creation accounts in Wisdom 8 and Psalm 104 while admitting that certain biblical texts do not reflect a genuine concern for Earth. I examined God's covenant with Noah in Genesis 9:12–13, which acknowledges that destruction is a reality that does not negate God's love for creation. I then drew on Johnson's insights about the Mystery that is present at the heart of the universe as an expression of loving self-surrender. Earth Link's theology acknowledged the centrality of the Christ event and identified Jesus as 'the embodiment of the connection, communion, unity of all created reality with the Sacred, an embodiment of God's nearness, and indeed of God's love.'[17] Jesus was recognised as one who 'showed his own and future generations that matter is good and that it embodies the Sacred.'[18] Taking Jesus as the norm for Earth and for humanity led naturally to Sallie McFague's map of what ecological Christianity might look like, and I quote it because it contains the seeds that come to fruition in this project:

• the insistence on justice to the oppressed, including nature, and

---

14 Ibid.
15 Keller, *On the Mystery,* p. 53.
16 Tinney, Ecology and Christian Faith.
17 Ibid.
18 Ibid.

the realisation that solidarity with the oppressed will result in cruciform living for the affluent

- the need to turn to the earth, respecting it and caring for it in local, ordinary, mundane ways
- the recognition that God is with us, embodied not only in Jesus of Nazareth, but in all of nature, thus uniting all creation and sanctifying bodily life
- the promise of a renewed creation through the hope of the resurrection, a promise that includes the whole cosmos and speaks to our ecological despair
- the appreciation of the intrinsic worth of all life forms, not just of human beings
- the acknowledgement that human salvation or wellbeing and nature's health are intrinsically connected.[19]

Earth Link's theology went on to acknowledge God's presence in the Spirit in the unfolding story of the universe, 'present in its autopoesis or self-emergence, present in its groaning, present intimately in love at all time.'[20] The loving presence of God among us does not remove suffering and death, but it enables us to draw on the disturbing message from the death and resurrection of Jesus that there are 'other sources of life, love and power than those being demonstrated by civil and religious leaders'[21] of his time and ours. This encounter with Mystery has the power to draw us forward into future fullness of being, even if this is beyond our imagination. With Catherine Keller, I was drawn to 'an exploration of the future as whirlwind, an exploration of the creative edges of chaos which brim with potential.'[22] With this rudimentary theological framework, Earth Link strained forward in hope towards an ultimate encounter with Mystery, but also in the hope that it was making a contribution to the God-Earth-human connection.

Earth Link is a faith-based community response to the ecological

19 McFague, 'Ecological Christianity', p. 33.
20 Tinney, *Ecology and Christian Faith.*
21 Ibid.
22 Ibid., drawing on Catherine Keller, 'No More Sea'.

crisis. It is a child of its time and reflects the language of the resources available to us in its formative and ongoing stages. We at Earth Link were very keen to be inclusive, and to use images and metaphors that were easily accessible to those who gathered around it. Our inner circle included Christians, post-Christians and proponents of Goddess religion. It is my experience that many people who are aware of environmental and other global concerns are caught between the classical and the evolutionary paradigms, and may be in a phase of 'new atheism' such as Ilia Delio speaks of.[23] Such people are not so averse to ecospirituality, and I think that Earth Link might have discovered something in developing it as a focus when we did.

I believe that our vision that the whole Earth community is worthy of our respect, reverence and care is an enduring one. Our reverence leads us to recognise the sacredness of Earth as the place of encounter with the Sacred/the Divine/God. Our respect is based on our recognition of the intrinsic value of the whole Earth community. Our care comes from our oneness with the Earth community and our kinship with its members, and is amplified by our sense of its sacredness. Our intention is that this vision will take us beyond treating Earth as an object for our gratification and exploitation.

The ecospirituality and ecotheology of Earth Link are intended to enable Heaven and Earth to embrace in the lives of those who live in our emerging universe. However, Earth Link's story is itself unfolding. This awareness led me to embark on a process of research through the writing of a doctoral thesis. From that research I have reconstituted the principles for Earth Link, integrating ecospirituality and ecotheology into a framework that builds on what we have developed to date. These principles are offered in detail in this book.

---

23 Delio, *Unbearable Wholeness*.

CHAPTER 3

When Heaven and Earth Embrace

A thesis is a literary form favoured by the academic community. Writing a thesis that bore the same title as this book afforded me the opportunity to access resources and supervisors who assisted me in getting the work to the standard required for a Doctorate. I am grateful to them and to Australian Catholic University for this opportunity. Rather than reproducing the thesis in this book[1], I will give an overview of that work and indicate how it has led to the development of enhanced principles for Earth Link, which I will present in full.

Practical theology

I chose the methodology of practical theology for the thesis, and hence for this book, because it is about exploring connections between Heaven and Earth, science and religion, spirit and matter, theory and practice, and theology and life. Don Browning, one of the pioneers of the discipline, sees it as a task of any theology to link practice and theory.[2] More specifically, in developing the notion of strategic practical theology as a movement within a framework that integrates descriptive theology, historical theology, and systematic theology, his intent is to underline the importance of integrating practice and theory, moving on to enhanced practice. He contrasts this with the movement from theory to practice that prevailed in systematic theology up to the 1980s. While systematic theology is predominantly interested in seeking meaning, practical theology begins with practice and seeks to improve it.

The interconnection between practice and theory, and life and theology, is also stressed by Miller-McLemore in the Introduction to the

---

1 Tinney, Heaven and Earth.
2 Browning, *Practical Theology*, p. 9.

Wiley-Black Companion to Practical Theology. She speaks of practical theology as a way of life, a method, a curriculum and a discipline, and stresses the importance of the first and the last of these if we need to be reminded that the ultimate aim of practical theology 'lies beyond disciplinary concerns in the pursuit of an embodied Christian faith.'[3] As she says later in that same introduction, '[P]ractical theology either has relevance for everyday faith and life, or it has no meaning at all.'[4]

According to Veling, practical theology recognises, along with Buber, that 'religious experience is not meant to "lift you out of the world", but to lead you into the world.'[5] It wants to 'keep our relationship with the world open, so that we are never quite "done" with things; rather, always undoing and redoing them, so that we can keep the doing happening, passionate, keen, expectant-never satisfied, never quite finished.'[6] He goes on to say that

> [p]ractical theology is an effort to always honour the appeal to human experience, drawing our attention to questions of history, culture, and society, urging us to respond to the real needs of our world, to the conditions of human existence, 'on earth.' This is perhaps what is meant by the word 'practical.' Yet it is practical *theology* – an effort to regain the transcendent appeal of God's word to humanity, an appeal that calls out to us and asks us to be people of God, people of faith, people of hope, people of justice and mercy – a people living and acting on earth, 'as it is in heaven.'[7]

In other words, practical theology honours religious experience in all its unfinishedness, even though it is about 'a certain reintegration of theology into the weave and fabric of human living, in which theology becomes a "practice" or a way of life.'[8] It is about learning in the doing rather than abstract theorising about a situation.

In emerging areas such as ecospirituality and ecotheology, there

---

3 Miller-McLemore, 'Contributions of Practical Theology', p. 5.
4 Ibid., p. 7.
5 Veling, *Practical Theology*, p. 9.
6 Ibid., p. 7.
7 Ibid., p. 18.
8 Ibid., p. 3.

needs to be room for a critical examination of traditional sources. I find Browning's work particularly helpful because he highlights that practical theology 'brings questions generated by the disruption of experience to its tradition-saturated ideals and practices.'[9] This process, for me, begins with a sense that classical approaches and understandings are indeed in serious need of a critical dialogue with contemporary scientific understandings about the emergence of the universe. But the reverse is also true. Contemporary world views need to be enriched by religious traditions. There are problems and possibilities here that need to be explored. Browning embraces this view in his definition of fundamental practical theology as 'critical reflection on the church's dialogue with Christian sources and other communities of experience and interpretation, with the aim of guiding its action towards social and individual transformation.'[10]

The intent of practical theological reflection, according to Darragh, is 'transformative practice'[11]. For me, the ethical response of 'transformative practice' is animated and sustained by spiritual practice, hence my interest in the embrace of Heaven and Earth. The goal of this project is environmentally responsible behaviour driven by both love and concern for our home, Planet Earth, and her inhabitants,[12] and by 'respect, reverence and care'[13] for them.

Method/process/craft

Considering both the presuppositions and the hazards of practical theology, Darragh summarises its method/process/craft as follows:

> Briefly, the process begins from the analysis of a particular context by researchers who are themselves part of that context. This analysis requires a description, normally non-theological, of an area of interest

---

9 Browning, *Practical Theology,* p. 11.
10 Ibid., p. 16.
11 Darragh, 'Practice of Practical Theology', p. 1.
12 The use of the feminine when referring to Earth is discouraged by feminist scholars, who see it as reinforcing the oppression of both women and Earth. I am choosing rather to align with ancient and contemporary traditions that honour Earth as Mother, seeing her as worthy of the greatest respect, reverence and care.
13 See the mission of Earth Link as stated on its website, www.earth-link.org.au.

within the contemporary context for the purpose of deciding on a significant ethical issue within that area of interest. The researchers need, in the course of this analysis, to pay attention to their own position within that context, and to take note of analyses that have been made in different but related contexts. Once a decision has been made on the contemporary issue to be investigated, the researchers formulate this issue into a question (a 'pivotal question') that can be addressed to Scripture and Christian Tradition. The researcher then reads the Christian sources in the light of that question, taking into account accepted principles of scriptural interpretation, in the hope of arriving at a response. If a scriptural and traditional response can be arrived at, the researcher seeks a contemporary application that will transform contemporary action in the original context.[14]

This essentially cyclical process is not unique to practical theology, as the see-judge-act process is intrinsic to most praxis methodologies.[15] Basically, a method is a map and not the route itself. It helps to have a map, but life is not usually that linear. It ducks and weaves and adjusts in the light of new learnings and insights. The action-research approach,[16] which is spiral in its concept, reminds us that the end is often the beginning of a new cycle. A major learning in writing the thesis was that there is a meaningful interconnection between ecotheological discourse and ecospiritual practice, and that this interconnection can in turn enhance the principles and practice of Earth Link.

The Sisters of Mercy regularly engage at various levels of the organisation in processes that follow the theological reflection cycle, as depicted in Figure 1. At a gathering of theologians in Burlingame, California, in 2007 the Sisters constructed an initial visual model. This model has since been refined as the basis of a global Mercy International Reflection Process focussed on listening to the cry of Earth and the cry of the poor. I will use this diagram as a template for the book.

---

14 Darragh, 'Practice of Practical Theology', p. 3.

15 Most praxis methodologies begin by seeing or describing the current reality, then judging or critiquing it against a set of values, after which plans for action are developed.

16 Action research is a praxis methodology that goes through the see-judge-act process, then begins it again to examine the effectiveness of the action.

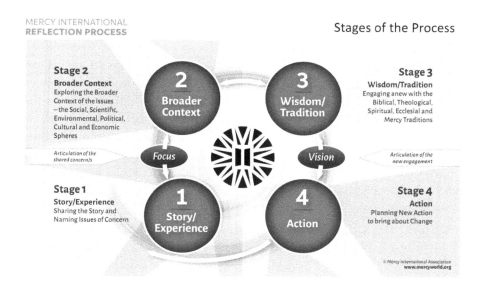

Figure 1: Mercy International Reflection Process - theological cycle[17]

This book is about bringing the craft of practical theology to bear on ecospirituality in an emerging universe, on the interconnectedness of God, humans and nature, on the embrace of Heaven and Earth. Wolfteich speaks specifically about a practical theology approach to the study of spirituality. She envisages a partnership between professionals from both fields using a process such as that outlined in Figure 1. Of value is her suggestion that practical theology can contribute to such a dialogue in a number of ways: thick description of spiritual practice that would uncover their implicit world views and theologies, access to the riches of the tradition out of which the spirituality emerges, the correlation of contemporary practice and the tradition such that past and present are brought into dialogue, and the enriching of formative processes in the light of the above.[18] These contributions enhance this research.

---

17 Mercy International Association, ©2018. Mercy International Reflection Process Guide Book, Adele Howard (Editor). Used by permission.
18 Wolfteich, 'Animating Questions', pp. 138–9.

Dialogue with ecospirituality

We begin this first dialogue by exploring the nature of ecospirituality, before considering Thomas Berry as an exemplar, given his status as a visionary leader in the field. There are academic descriptions of ecospirituality, but I believe that Gerard Manley Hopkins gets to the point when he remarks, after describing the harvest field, that 'these things, these things were here and but the beholder wanting'.[19] In the ecospiritual experience, the beholder 'sees' what is 'here' with new eyes, the eyes of 'rapturous love'. This loving presence is here and 'but the beholder wanting'. Ecospirituality is an opportunity for Heaven and Earth to embrace in the depth of our being.

Spirituality does not belong solely in the domain of formal religion. I distinguish between the Gaians with their sense of animate Earth, those who experience Earth as sacred (whether in a theist or non-theist sense), and those with what Rasmussen calls an 'Earth-honoring faith', which is panentheistic in that God is in nature but not contained by it. Earth Link belongs in this last category and grounds its ecospirituality in the Christian tradition. Christian spirituality is not a case of 'applied' theology, but instead involves a dialogue between spiritual experience and the insights of the tradition, in order to enliven experience.

Thomas Berry – visionary leader

Thomas Berry epitomises in so many ways the embrace of Heaven and Earth. Core among his suggestions as to what is needed in order to redress the crises of our times is a spirituality that is deeply grounded in the emerging universe. Bron Taylor described Thomas Berry's practice as an example of dark green religion 'in which nature is sacred, has intrinsic value, and is therefore due reverent care.'[20]

Berry lived from 1914 to 2009. He was a cultural historian by training, as well as a Passionist priest. He was director of Fordham's graduate programme in the history of religions from 1955 to 1979. In 1970 he founded the Centre for Religious Research in Riverdale, New York, and was its director until 1987. It was during this last period

---

19 Gerard Manley Hopkins, *Hurrahing in Harvest,* https://www.bartleby.com/122/14.html.
20 Taylor, *Dark Green Religion,* p. 10.

that he began to lecture widely on the intersection of cultural, spiritual and ecological issues. His writings on ecospirituality and the human relationship with the Earth began in 1988. He wrote prolifically, and *Dream of the Earth* and *The Universe Story,* which he co-authored with Brian Swimme, could now be counted among the classics. He authored many, many essays and gave many talks; these are being gathered into volumes of his writings and published posthumously.

Berry was intensely conscious of planetary degradation. He saw its root cause in human alienation from the natural environment. Berry summed up his core message by saying, 'The Great Work, now, as we move into a new millennium, is to carry out the transition from a period of human devastation of the Earth to a period when humans would be present to the planet in a mutually beneficial manner.'[21] Berry has several key insights into how our alienation from nature could be addressed. The first is that we need to be aware of our place in the emerging universe. Our insertion into a time-developmental or cosmological perspective enables us to recognise the differentiation of all individuals, their unique inner life and identity, and their capacity for communal bonding. Berry calls these the principles of differentiation, subjectivity and communion.[22] For him this is not just a story about matter. The human participation in the dream of the Earth is 'an experience wherein human consciousness awakens to the grandeur and sacred quality of the Earth process.'[23]

His message is not confined to individuals but extends to legal, economic, political and religious systems. He was a pioneer, calling these systems to account with his analysis and his exhortations. Berry could live with ambiguity when it came to the church. He was not dismissive of traditional religion but recognised that it could not go ahead on its own, and he was aware that society could not go ahead without it. He did, however, acknowledge that a beginning was being made to bring religions into the ecological age.

There are glimpses into the moments of insight that stirred him on. In *The Great Work,* he referred to an experience in his childhood that

---

21 Berry, *Great Work*, p. 3.
22 Ibid., p. 162.
23 Ibid., p. 165.

became normative for him. His magic moment in a flowering meadow one morning in May in his childhood is one of these. It leads him to say, 'That is good in economics which fosters the natural processes of this meadow. So in jurisprudence, law, and political affairs – what is good recognizes the rights of this meadow and the creek and the woodlands to exist and flourish in the ever-renewing seasonal expression.'[24]

Berry's spirituality built on this transformative experience of the meadow, which enabled him to call others to moments 'akin to ecstasy'.[25] His is an intimate experience of the natural world, an experience of the natural world as subject rather than as an object to be exploited.[26] It is even more – it is recognition of the numinous qualities of Earth.[27]

To Berry, Earth was sacred. To him, the new story of the universe is now our sacred story. He urged us to recover the book of nature as the primordial source of God's revelation. While Berry draws on the wisdom of his patron Thomas Aquinas and on Teilhard de Chardin, it is not to him that I look for an extended theological reflection to inform Earth Link's ecospirituality. Resonances of ecospirituality are found in other writers; these can amplify our understanding and influence our praxis. Such writers establish ecospirituality as

- cosmic in scope: connecting spirit with all of life
- attentive to the sacred inner dimensions of nature
- open to transformative, mystical encounters with nature
- motivated for justice for the whole Earth community.

These themes influence strongly the enhanced principles for Earth Link that flow from the dialogues I conduct with ecospirituality and ecotheology. It is to the ecotheology of Elizabeth Johnson and Pope Francis that I now turn.

Dialogue with ecotheology

In choosing to dialogue with the work of Elizabeth Johnson, I am

---

24 Ibid., pp. 12–13.
25 Berry, *Sacred Universe*, p. 132.
26 Swimme & Berry, *Universe Story*, p. 243.
27 Berry, *Sacred Universe*, p. 73.

engaging with someone who recognises that there are three main pillars of theology: nature, God and humanity. The embrace of these three is the focus of the book in its exploration of how to engage spiritually in an emerging universe. Where relevant, I refer to *Laudato Si,* the 2015 encyclical of Pope Francis, which Johnson herself describes as 'the most important encyclical ever written in the history of the Catholic Church.'[28]

There are seeds of an ecotheology in Johnson's work prior to the publication of *Ask the Beasts* in 2014. By the time she gave the Madeleva lecture in 1993, her goal was a 'flourishing human community on a thriving earth.' Her Presidential Address to the Catholic Theological Society of America in 1996 was entitled 'Turn to the Heavens and the Earth: Retrieval of the Cosmos in Theology', and she exhorted her colleagues 'to reclaim the cosmos as a theme in theology so that access to the fullness of theological revelation might be restored.'[29] Johnson's most controversial book, *Quest for the Living God,* has a chapter entitled 'Creator Spirit in the Evolving World'. In it, Johnson begins with the context where encounters with nature are accompanied by both wonder and waste. Yet in those experiences, she recognises that 'people of faith are rediscovering an ancient theme, namely, the presence and action of the creative spirit of God throughout the natural world.'[30] Johnson acknowledges that recent theology has not been very helpful in making meaning of this faith experience, and she reiterates her diagnosis that this is due to shortcomings in the theology of the Spirit. For Johnson, '[a]ttending to the idea of the Creator Spirit brings to the fore the belief that the presence and activity of God pervade the world and therefore the natural world is the dwelling place of God.'[31] Divine presence is continuous, cruciform, and abides in the mode of promise. In summary she says:

> Ecological theology proposes that the Creator Spirit dwells at the heart of the natural world, graciously energizing its evolution from

28 Johnson, 'Pyramid to Circle', pp. 479–87.
29 Johnson, 'Turn to the Heavens', p. 5.
30 Johnson, *Quest*, p. 182.
31 Ibid., pp. 187–8.

within, compassionately holding all creatures in their finitude and death, and drawing the world forward toward an unimaginable future. Throughout the vast sweep of cosmic and biological evolution, the Spirit embraces the material root of life and its endless new potential, empowering the cosmic process from within. The universe, in turn, is self-organising and self-transcending, energized from the spiralling galaxies to the double helix of the DNA molecule by the dance of divine vivifying power.[32]

Hers is a response to her diagnosis that the malaise experienced by those at the margins and by Earth is the result of ignoring the theology of the Spirit, and hence her interest in its retrieval. Several ecological themes are explored for their theological significance in this book and developed more fully in *Ask the Beasts*. They include suffering and death in nature, theological models that enable her to conclude that 'God makes the world … by empowering the world to make itself,'[33] and the inadequacy of static or interventionist theologies in the face of newer understandings of the cosmos. She points to the challenges that ensue, whether these be theological retrieval, mystical engagement or prophetic action.

Johnsons' theology as presented in *Quest for the Living God* resulted in her being the subject of a doctrinal investigation by the United States Conference of Catholic Bishops (USCCB). I explored the clash of theologies and philosophical assumptions that was manifest in this unfortunate encounter, and decided in favour of Johnson. I was influenced by this assessment of the situation by Richard Galliardetz:

> The current magisterial tendency to rush to doctrinal judgments with every new theological foray forgets Newman's important insight: divine truth, emerges only slowly, patiently, and always with a certain tentativeness. The work of theology is akin to the ministrations of a midwife; it is the work of theology to assist patiently in the birthing of God's Word in our time. By contrast, the rush to doctrinal judgment is not unlike the frantic father wishing to hasten the birthing process

---

32 Ibid., p. 191.
33 Ibid., p. 193.

even if it places mother and child at risk.[34]

Ask the Beasts

I chose Johnson's book *Ask the Beasts* as the basis of an in-depth study of contemporary ecotheology.

Johnson is the major dialogue partner in this theological reflection. Thomas Berry was influential in the setting up of Earth Link in 2000, but ecotheology was less developed in those days. Earth Link can now reap the benefits of developments that have taken place in this field since its inception.

Johnson's focal question is 'What is the theological meaning of the natural world of life?'[35] She addresses her question by means of a dialogue between Charles Darwin, especially through his work *On the Origin of Species,* and the Christian creedal statement, the Nicene Creed. While Johnson stands firmly within the Christian tradition, and alongside other ecotheologians, her approach is distinctive. She draws on Job for her starting point and operative approach:

> 'Ask the beasts and they will teach you,' says Job (Job 12:7); speak to the birds of the air, the plants of the earth, and the fish of the sea and they will instruct you. On the face of it, this seems a simple thing to do: consult the creatures of the earth and listen to the religious wisdom they impart. Given theology's longstanding preoccupation with the human drama, however … the invitation to consult the plants and animals harbors the demand for a subtle change of method. […] The effort to approach other species with concentrated attention to their story in all its struggle and delight creates an important shift in perspective. […] The focus has to shift to those who have been silenced, so that they are seen as of central importance in themselves. In a similar manner, the nascent field of ecological theology asks that we give careful consideration to the natural world *in its own right* as an irreplaceable element in the theological project.[36]

---

34 Galliardetz, *When the Magisterium Intervenes*, p. 251.

35 Johnson, *Ask the Beasts,* p. xiv.

36 Ibid., p. xv.

Johnson enters into the science-religion dialogue by examining Darwin's theory of the evolution of the species, then moves on from science to unfold the relationship between the evolving world and God understood as Trinity. She concludes by acknowledging the harmful impact of humans, before locating them within the community of life and as members of the community of creation, all of which is in relationship with God. This is the basis of the responsibility that comes from kinship in those communities, and which impels us to action for Earth's welfare. This theology is intended to benefit the natural world in a time of crisis by exploring how 'love of the natural world is an intrinsic act of faith in God, to practical and critical effect.'[37]

In a discussion at the meeting of the Catholic Theological Society of America in June 2014, Johnson summed up her book by saying:

> In a nutshell, this project consists of swivelling our gaze away from the fascinating theological mirror that reflects our own faces, and toward looking out the window at the evolving natural world now under threat. Beholding creation on its own, non-anthropocentric terms, the book aims to show how loving the Earth arises as an intrinsic part of faith in God, rather than just an add-on.[38]

Johnson begins by asking the beasts – in this case, the species studied by Charles Darwin. She is strong in her praise of his obvious love of nature, and of the sometimes lyrical style of his writing. She summarises his core insights thus:

> In face of the widespread scientific and religious assumption that species come into being independent of each other by separate acts of a divine Creator, and the view that they remain immutably themselves throughout their existence, *On the Origin of Species* is one long argument that species are in motion, coming into being from previous species by a process that can be explained naturally, without appeal to a supernatural cause.[39]

The species have told her that they are interconnected and evolving, but there is more. In words reminiscent of her focal question, Johnson

---

37 Ibid., p. xviii.
38 Johnson, *Pyramid to Circle*, p. 479.
39 Johnson, *Ask the Beasts,* p. 27.

reminds us that 'the theory of evolution is theologically consequential' and asks, 'How shall we speak of the overflowing love of the creating, redeeming, re-creating God of life in view of evolution?'[40]

In *Ask the Beasts,* Johnson transitions from Darwin's science into theology by looking at nature through the eyes of Job: 'Ask the animals, and they will teach you …[T]he hand of the Lord has done this[.] In his hand is the life of every living thing' (Job 12:7, 9–10).

She keeps her eyes on Darwin's image of the tangled bank of plants, birds, insects, worms and damp earth, and their 'elaborately constructed forms, so different from each other and dependent on each other in so complex a manner,'[41] but now she enters into the world of religious affirmation. She sees the tangled bank as 'created', a declaration that does not belong to science, but is 'the expression of a basic trust that the universe has an ultimately transcendent origin, support and goal which renders it profoundly meaningful.'[42] Its vitality is a gift from God's hands, a gift which brings it to reality and sustains it in existence, and without which it would not exist at all. Johnson draws on classical creation theology, which encompasses 'original creation in the beginning, continuous creation in the present here and now, and new creation at the redeemed end time.'[43] In the Judeo-Christian tradition, plants and animals have a reference point beyond themselves in the 'overflowing generosity of the incomprehensible God who freely shares life with the world.'[44] God's presence is there as origin, source and goal. It is an ongoing and dynamic presence, and Johnson borrows the metaphor of the singer keeping the song in existence at all times to give the sense of immediate and sustained presence.[45] God's presence is traditionally known as the Spirit who also 'continues to draw the world to an unpredictable future, pervaded by a radical promise at the ultimate end of time.'[46]

Johnson argues that it best suits her dialogue with Darwin to begin by

---

40 Ibid., p. 121.
41 Darwin, *Annotated Origin*, p. 489.
42 Johnson, *Ask the Beasts,* p. 123.
43 Ibid.
44 Ibid., p. 124.
45 McCabe, *God, Christ and Us*, p. 103.
46 Johnson, *Ask the Beasts*, p. 124.

exploring *creatio continuo:* to begin with the here and now, with God's Spirit as 'ground and bearer of all evolutionary life.'[47] In Johnson's words, 'Earth is a physical place of extravagant dynamism that bodies forth the gracious presence of God. In its own way it is a sacrament and a revelation.'[48] Its materiality embodies 'the active presence of the holy Giver of life,'[49] an understanding that is acted out in the sacraments of the Church, itself 'constituted as a sacrament of Christ's presence for the world.'[50] Matter does reveal, in the biblical sense of teaching us about God.

Johnson does not stop here, as she is cognisant of the impact of evolutionary emergence and the theological questions that arise as one comes to terms with the unfolding nature of reality driven by capacities that are inherent in it. In response to such concerns she considers ways of approaching God's action in the processes of evolutionary emergence. The constructive new theological interpretation that she offers is that 'God's creative activity brings into being a universe endowed with the innate capacity to evolve by the operating of its own natural powers, making it a free partner in its own creation,'[51] a model based on love rather than power over. This is no deistic God who creates and then leaves the world to its own devices. This is no monarchical God who manages every detail. It is rather that the Giver of life 'freely and generously invests nature with the power to organise itself and emerge into ever-new, more complex forms, and to do so according to its own ways of operating.'[52] It is always being called forward to newness while being empowered from within.

The evolutionary story is marked by breakdown as well as creativity. Darwin's findings remind us that evolution comes at a cost. His theory of descent with modification raised new questions about the value and inevitability of suffering and death. Johnson addresses these questions in the chapter where she 'links the Creator Spirit present and active in

47 Ibid.
48 Ibid., p. 150.
49 Ibid., p. 151.
50 Ibid.
51 Ibid., p. 155.
52 Ibid., p. 156.

the world with the love of God made known in the death and resurrection of Christ, beginning with the groaning and then moving to hope.'[53]

Johnson theologically frames the conversation that is needed about this reality. As she did in her earlier writings with the question of human suffering, Johnson rejects theodicy that seeks to 'construct a rational defense of God's goodness and power in a world where evil occurs'.[54] Such justification creates a situation of passive acceptance of enormous suffering, and of the political and economic situations that can give rise to it.[55] Instead, she seeks 'a theological inquiry that takes the evolutionary function of affliction at face value and seeks to reflect on its workings in view of the God of Love made known in revelation.'[56]

Johnson indicates that 'the most fundamental move theology can make … is to affirm the compassionate presence of God in the midst of the shocking enormity of pain and death.'[57] In the Hebrew Scriptures, God is a God of pathos, variously delighted with Creation and full of lament for the devastation of land and people. The Christian tradition is built on the life, suffering, death and resurrection of Jesus Christ, one who entered into the beauty and the pain of an earthly existence to the point of dying a tortuous death. However, in the experience of encountering the resurrected Christ, the tradition came to the belief that the death of Jesus was not the end, but indeed part of a new creation which forms the basis of hope.

Johnson explores the connection between the groaning of creation and the God of Love, and draws on her earlier exploration of God present in Jesus Christ in solidarity in the midst of suffering. Johnson follows the Danish theologian Niels Gregersen, who coined the phrase 'deep incarnation', which Johnson explains as 'this radical divine reach through human flesh all the way down into the very tissue of biological existence with its growth and decay, joined with the wide processes of

---

53 Ibid., p. 184.
54 Ibid., p. 187.
55 Johnson would probably agree with Delio's suggestion that the real theodicy question is 'not why God allows bad things to happen to good people, but why we abandon God in the face of suffering'. See Delio, *Unbearable Wholeness*, p. 83.
56 Johnson, *Ask the Beasts,* p. 187.
57 Johnson, *Ask the Beasts*, p. 191.

evolving nature that beget and sustain life.'[58] Creator and created are radically connected in the person of Jesus. Jesus in turn is radically connected to the stuff of the universe. As Gregersen says, Jesus bears 'the signature of the supernovas and the geology and life story of the Earth.'[59] Jesus is bonded, not only to humans whose form he took, but to the universe that gave rise to his matter. In the incarnation, Jesus assumes the universe, not just human flesh.

Jesus paid a very high price for his commitment to the liberating love of all creation. He suffered an excruciating death at the hands of his captors. Classical theology has long sought to hold together an understanding of the human and divine natures in the person of Jesus, which they have expressed as a hypostatic union. Johnson draws on the more contemporary theology of Walter Kasper, who points to the unexpected, shocking death of Jesus on the cross as 'the unsurpassable self-definition of God.'[60] This suffering, freely chosen, is an expression of love. For the early Christians and for us, his death was and is not the end. Johnson takes Gregersen's deep incarnation into deep resurrection, which she says 'extends the risen Christ's affiliation to the whole natural world.'[61] The risen Christ takes corporeality in all its manifestations into the heart of God.

Johnson began with continuing creation (*creatio continuo*) because she said that this best suited her dialogue with Darwin and the beasts. Now she looks at creation 'beyond time'[62], namely, in the beginning (*creatio originalis*) and in the new creation (*creatio nova*). She adopts a 'hermeneutic of the present experience of grace'[63] to link the God of the beginning with the God of the end times, and to make the case for the action of Holy Mystery in creating life and working towards its ultimate fulfilment beyond death. She revisits traditional understandings of creation and affirms that reality, as created, is God's good gift, given in love. She considers some of the scientific scenarios about the future of

---

58 Ibid., p. 196.
59 Gregersen, *Cross of Christ*, pp. 192–207.
60 Kasper, *God of Jesus Christ*, p. 194.
61 Johnson, *Ask the Beasts, p.* 208.
62 Ibid., p. 212.
63 Ibid., p. 213.

the universe, and ways in which religious thinking around eschatology has developed. She underlines that 'what such speech does do is affirm the core conviction that all of reality exists within the embrace of God's gracious love, and that it is going toward a fulfilment yet to come.'[64] Just as the creation story places reality in God's hands, the end times are about a return to the fullness of the love that originated, sustained and will draw it into wholeness beyond its finite existence.

To this point, her focus has been on the 'theological meaning of the natural world of life'.[65] Her gaze has been on 'those who have been silenced.'[66] It is now time to look at those who have done the silencing and revisit their relationship to the community of life, along with the ethical responsibilities that flow from this relationship.

The title of Johnson's chapter 'Enter the Humans' points to both the continuity and the discontinuity between the species she has been considering thus far and the human species. Our materiality is continuous with the stuff of stardust, and 'we humans ... share a genetic heritage with every other species of the tree of life, a biological kinship encoded in each cell of our body.'[67] Our distinguishing capacities of 'self-reflective consciousness and freedom'[68], or of 'mind and will', to use the classical terms, were new capacities for the universe. This, of course, is consistent with the world view that is the vision of this book, and for which I am using the symbol of the Celtic triquetra.

Johnson is strong in her description of negative human impacts on the community of life. When she analyses the current state of environmental destruction she suggests that '[i]f human beings were to wake up to the grandeur of the dying world, fall in love with life and change their behaviour to protect it, much of the current dying off could be slowly brought under control.'[69]

It is a short step from there for Johnson to add her voice to the call by recent Popes for ecological conversion, which she describes as 'falling

64 Ibid., p. 221.
65 Ibid., p. xv.
66 Ibid.
67 Ibid., p. 237.
68 Ibid., p. 236.
69 Ibid., p. 253.

in love with earth as an inherently valuable, living community in which we participate, and bending every effort to be creatively faithful to its well-being, in tune with the living God who brought it into being and cherishes it with unconditional love.'[70]

Johnson acknowledges the impact of the paradigm of dominion, attributed to the Judaeo-Christian tradition in its origins and flowing over into societal norms. Then she develops the very rich, new paradigm of the community of creation. This is the high point of her work, in my opinion. The theological construct of the community of creation is founded on the belief that 'all beings are in fact creatures, sustained in life by the Creator of all that is.' This is the case for humankind and other species, and this commonality before God is stronger than their differences. In their kinship all are 'grounded in absolute, universal reliance on the living God for the breath of life.'[71] This pattern of relationship, which locates us humans alongside other creatures and in relationship with God, gives a new impetus for ethical behaviour based on that new relationality with one another and the wider whole, and can supersede notions of dominion. This clarified relationship is of great significance for the enhanced principles and practice for Earth Link.

Affirmations and challenges

Earth Link was the focus of the thesis, just as it is of this book. The experience of Earth Link was taken into dialogue with Thomas Berry and some other writers in the area of ecospirituality, with Elizabeth Johnson in the area of ecotheology, and with the encyclical *Laudato Si* of Pope Francis that was promulgated in 2015. Earth Link was affirmed by Sandra Schneiders' approach to spirituality as 'not only a complexification of the holistic approach to the human subject of religious experience, but a heightened awareness of the dimensions and influence of "place" and "space" (both inner and outer), globalization, ecological crises, the validity of religious experience outside one's own tradition, scientific developments, and cultural currents.'[72] Earth Link's

---

70 Ibid., p. 259.
71 Ibid., p. 268, is the source of the above quotes.
72 Schneider, *Christian Spirituality*, p. 10.

approach is both affirmed and challenged by the need to be aware of its own tradition as it is developing, and also to be cognisant of other spiritual traditions with which it enters into partnership for the well-being of Planet Earth and the cosmos. Earth Link is also mindful of Schneiders' warning:

> The challenge for those who approach the study of spirituality from the more anthropological perspective is to keep the specifically Christian character of the discipline in focus and to resist the postmodern lure of universal relativism, nihilistic deconstructionism, rejection of all tradition and authority, and suspicion of personal commitment.[73]

Earth Link is affirmed by revisiting the work of Thomas Berry and by having access to his previously unpublished material, even as we now recognise the need to further integrate a theological framework, a task that Berry only lightly touched on. Other writers in the area who resonate with the approach of Berry present between them an understanding of ecospirituality as cosmic in scope, attentive to the Sacred, open to transformative, mystical encounters with nature, and providing a driving force for action for justice for the whole Earth community. Earth Link is affirmed by many of these developments even as if marks out its own particular domain.

One area in which the approach of Earth Link was less developed was that of ecotheology, and there have been significant developments in that field since Earth Link began in 2000. Elizabeth Johnson published her major work on ecotheology in 2014, just as I was entering into the theological part of the thesis. Her work is comprehensive and accessible in its often poetic style. It challenges Earth Link to develop a comprehensive Trinitarian ecotheology cognisant of findings in contemporary science. Within that, Earth Link is being led, above all, to a new understanding of the pathos of God, a God who, in the biblical tradition, is moved by the plight of Earth and her people. God entered into solidarity with all created reality in the person of Jesus, whose life, death and resurrection point towards the possibility of fullness of life and love in God forever. Earth Link is also challenged to take forward

---

73 Ibid., pp. 11 & 14.

its notion of the 'Earth community' into the theological notion of the 'community of creation'.

Pope Francis' release in 2015 of *Laudato Si* significantly influenced my work. References to it are integrated into passages on Elizabeth Johnson.

These affirmations and challenges were taken forward into the review of the vision, mission and principles of Earth Link, which form the bulk of this book. They are presented here, with explications drawn from the thesis. In addition I offer suggestions for individuals and groups who are seeking a framework for their own praxis, understood as informed practice.

This book is written in an era when the only constant is change. Human abuse of the environment and the reality of climate change continue to loom large on the public agenda. A continuing response to these realities is needed by all. Those from religious and other traditions who share Earth Link's vision of a world where there is 'respect, reverence and care for the whole Earth community' have an added responsibility to work for its enhancement. I believe this book can facilitate that much-needed response.

CHAPTER 4

Towards deep bonding

Earth Link aims to make a contribution to environmentally responsible behaviour driven by both love and concern for our home, Planet Earth, and her inhabitants, and by 'respect, reverence and care'[1] for them. Its hoped-for outcome is ecological conversion understood as 'falling in love with earth as an inherently valuable, living community in which we participate, and bending every effort to be creatively faithful to its well-being, in tune with the living God who brought it into being and cherishes it with unconditional love.'[2] I chose the craft of practical theology because its ultimate aim 'lies … in the pursuit of an embodied Christian faith.'[3] For Darragh and for this book, practical theology is about 'transformative practice'.[4] I have worked towards these outcomes by developing an enhanced set of principles for the Earth Link community. In this way, it will be possible for the dialogues in this research to enhance the practice of Earth Link. These new principles will have the potential to take the nineteen-year history and experience of Earth Link to a new place.

As stated at the beginning, I argue that it is possible for us to engage spiritually in an emerging universe if we have a vision of the embrace of Heaven and Earth that is informed by contemporary science, and if we underpin that with an ecotheology that recognises Heaven and Earth as interconnected while respecting their differences, at the same time cultivating an open ecospiritual praxis that is attentive to, and aware of, divine presence in all that is. This integrated approach is a feature of the enhanced principles.

---

1 See the mission of Earth Link as stated on its website: www.earth-link.org.au

2 Johnson, *Ask the Beasts,* p. 259.

3 Miller-McLemore, *Contributions of Practical Theology,* p. 5.

4 Darragh, *Practice of Practical Theology*, p. 1.

Initially Earth Link needs to revisit its vision and mission, modifying them where needed in the light of the dialogues in this book. Its broad vision is of a world where there is 'respect, reverence and care for the whole Earth community,' a vision that enshrines the attitudes of care and concern, plus the sense of Earth as sacred and having an inner dimension, which means it is deserving of reverence. This vision embodies a desired outcome for planet and people, especially at a time of widespread poverty and degradation of the planet. It could be enhanced to become a vision of a world where there is 'respect, reverence and care for the whole Earth community, held as it is in the embrace of the Divine.' This amendment adds the theological insight that Heaven and Earth embrace, with due recognition of their distinctiveness, and without any assumption of their equality. However, after due consideration I now prefer that we stay with the original, as it is more catchy as a statement of vision. It would not really be enhanced by the inclusion of a rationale.

Earth Link's mission needs to be modified from facilitating 'deep bonding with the Earth community' to become the facilitation of 'deep bonding within the community of creation.' Thus modified, the bonding is with the whole Earth community and with God as part of the community of creation. Earth Link continues to believe that deep bonding has the potential to be an encounter with Holy Mystery that leads us to ethical behaviour and practices. The rewording of the mission is a further acknowledgement of the embrace of Heaven and Earth, an embrace that allows each component 'to stand in its own difference, but encompassed by a wider whole that affects their interrelatedness.'[5]

Earth Link continues to 'resource, reflect and act'[6] towards the achievement of its vision and mission, with particular focus on engaging spiritually in an emerging universe. There is an assumption based on experience that spirituality[7] is a key component in nurturing a vision

---

5 Johnson, *Ask the Beasts*, p. 269.

6 This strategy is on the Earth Link website: www.earth-link.org.au

7 In this context spirituality is understood as 'a quest to deepen, renew, or tap into the most profound insights of traditional religions, as well as a word that consecrates otherwise secular endeavours such as psychotherapy, political and environmental activism, and one's lifestyle and vocational choices.' See Taylor, *Dark Green Religion*, p. 3.

of the embrace of Heaven and Earth, and moving people to respectful, careful and reverent attitudes and action within the Earth community.

The existing ecospirituality principles of Earth Link have proved invaluable as a framework for education and reflection about the nature and practice of ecospirituality. The document 'Ecology and Christian Faith' further develops the theological underpinnings within the Christian tradition. I now offer five enhanced principles integrating ecospirituality, ecotheology and environmental ethics, which could be shared within and beyond Christian groups. They build on the earlier principles but go beyond them.

Earth Link now invites you to

1. recognise that the universe is a dynamic entity within which all is interconnected yet distinct.
2. cultivate an open, attentive and receptive attitude in order to enter into transformative, mystical encounters
3. acknowledge the indwelling presence of the Spirit in Earth and cosmos
4. acknowledge that in Jesus Christ the God of love embraces the cosmos in healing solidarity
5. live in right relations within the community of creation.

These enhanced principles reflect a cognisance of the planetary crisis, and of the need to adopt a world view, spirituality, theology and ethical framework that work to redress it. Principle 1 is about the world view, which follows on from insights derived from contemporary science indicating that the universe is evolving according to a dynamism within it, moving it toward the emergence of greater complexity and consciousness. This world view needs to situate humans within their evolutionary context, a world view dubbed anthropocosmic by Tucker and Grim.[8] It also needs to point to the necessity of ethical action to redress its negative effects. Principle 2 builds on existing Earth Link principles as they relate to underlying ecospiritual attitudes and practices. The

---

8 The term 'anthropocosmic' refers to 'a view of the human as having arisen from cosmological and ecological processes which orient humans in the universe and ground them in nature'. See Grim and Tucker , *Ecology and Religion,* pp. 43–4.

encounter with sacred Earth is interpreted out of a specifically Christian framework in Principles 3–5, with special reference to the continuing presence of the Spirit and God's solidarity with, and promise for, the Earth community, especially as manifested in the life, suffering, death and resurrection of Jesus Christ. The last principle locates humans within the community of creation and underlines the responsibility for created reality which flows from that. In these enhanced principles I draw on the work of the theorists who were the dialogue partners in this research, with particular reference to Thomas Berry and those of his tradition, and on the ecotheological work of Elizabeth Johnson, whose comprehensive Trinitarian theology is responsive to the crisis facing Earth and cosmos; and I am cognisant of the findings of contemporary science, especially the work of Charles Darwin. Where relevant, I make reference to *Laudato Si,* the encyclical of Pope Francis about care of our common home. I do not bring these theorists into dialogue with each other, as they basically complement each other, and together they can contribute to this integrated statement of principles for Earth Link.

Principle 1: The universe as a dynamic entity

Recognise that the universe is a dynamic entity within which all is interconnected yet distinct.

The universe that is deserving of our respect, reverence and care is a dynamic entity, and we need a world view that acknowledges its evolutionary emergence. We need to transition from the classical or traditional paradigm with some of its negative consequences to the embrace of a contemporary, anthropocosmic world view that addresses the interconnectedness of all in this dynamic universe.

As noted in this book, the need for transition is urgent. Johnson has described the current treatment of the environment as ecocide.[9] She notes that our Blue Planet is in jeopardy, and action is needed for its survival. She also suggests that 'ravaging of people and ravaging

---

9 Johnson, *Women, Earth*, p. 8.

of the land'[10] go hand in hand. Pope Francis speaks about the cries of Mother Earth. He details the symptoms of the current ecological crisis as manifest in pollution and climate change, issues around water, loss of biodiversity, changes in the quality of human life, the breakdown of society and global inequality.[11] Earth Link agrees with him that 'we need only take a frank look at the facts to see that our common home is falling into serious disrepair.'[12] Humans have much to account for in the current state of affairs.

Brian Swimme reminded us that humans are now a planetary power[13], a position that is reinforced by a group of contemporary scientists who are quantifying the safe operating space for nine major planetary systems. They point out that in the current Anthropocene epoch that we are entering, 'human activities now rival global geophysical processes,' and they warn that we humans 'need to fundamentally alter our relationship with the planet we inhabit.' The diagram from the Swedish Academy of Sciences used later in this chapter provides an assessment of the state of those systems and can be taken forward into the exposition of this principle.

Some writers indicate the choices that confront us if we take seriously the 'great turning' that Joanna Macy calls for. The transition Thomas Berry advocates is a choice between a technocratic era, dominated by technological development and fixes[14], and an ecozoic era, a period when 'humans will be present to the planet as participating members of the comprehensive Earth community.'[15] This is also the vision of Earth Link.

Pope Francis looks below the symptoms of the current ecological crisis to an analysis of its causes and points towards what he calls the dominant technocratic paradigm. He acknowledges the benefits that have come from technological knowledge but recognises that its basic premise is flawed. He says that '[t]his paradigm exalts the concept of

---

10 Johnson, *Ask the Beasts,* p. 6.

11 Pope Francis, *Laudato Si,* pp. 20–60.

12 Ibid., p. 61.

13 Swimme, *Powers of the Universe.*

14 See Hamilton, *Earth Masters*.

15 Berry, *Great Work,* p. 8.

a subject who, using logical and rational procedures, progressively approaches and gains control over an external object.'[16] Berry has consistently diagnosed that a source of the devastation of the planet lies in our treatment of the natural world as object to be exploited, and our subsequent disconnection from it.[17] Johnson points to the world views of hierarchical and Cartesian dualism underlying this mindset.[18] Originating in Hellenistic dualism, spirit and matter were seen as polar opposites, and their differences were stressed. As a consequence, soul was seen as superior to body, and humans were superior to the rest of the created order. In a patriarchal order, it was a short step to seeing men as superior to women. With Descartes, the human mind was differentiated from matter and superior to it. According to this world view, matter is subordinate to human need or want, even if this view is tempered by notions of responsible stewardship.

Those who recognise the limits of the classical paradigm, including Earth Link, go beyond describing the problems to promoting or embracing a world view that addresses them. In response to what Barbour refers to as the *contemporary period*, Pope Francis promotes an integral vision in which there is 'a relationship existing between nature and the society which lives in it,'[19] and Tucker and Grim develop an anthropocosmic view that 'orient[s] humans in the universe and ground[s] them in nature.'[20]

Ilia Delio calls the emerging world view *evolutionary*. Drawing on the Teilhardian scholar Ewert Cousins, Delio describes our times as a second axial period that is 'communal, global, ecological and cosmic.'[21] A mechanistic, static world view gives way to one that is organic, relational and conscious of the place of humans in an unfolding cosmos. Contemporary sciences have made major contributions to the understanding that the universe is a dynamic entity within which all is interconnected yet distinct, and some of these influences are noted below.

16 Pope Francis, *Laudato Si,* p. 106.
17 Swimme & Berry , *Universe Story*, p. 243.
18 Johnson, *Ask the Beasts,* p. 126.
19 Pope Francis, *Laudato Si,* p. 39.
20 See Grim & Tucker , *Ecology and Religion*, pp. 43–4.
21 Delio, *Christ in Evolution,* p. 28.

Influences on contemporary world views

The twentieth century has seen some extraordinary discoveries. Notable among these are developments in cosmology, Darwin's theory of evolution and the emergence of quantum science. Debates rage as theology either denies, defends, or responds creatively to these discoveries. Because this is the context in which we are living into these principles, consideration needs to be given to some of the developments that are giving rise to the contemporary evolutionary world view and calling forth new spiritual and religious responses such as that of Earth Link. The broadest context is cosmology, followed by evolutionary emergence.

Our place in the cosmos

Cosmology[22] has opened up vistas for understanding our place in the universe, which allows for a much bigger canvas for the human story. A cosmological perspective enables us to locate Planet Earth within the context of the Milky Way galaxy and the sandstorm of galaxies beyond that. Current projects in scientific cosmology are even more expansive. They are directed at understanding the 'outer' space that is beyond the galaxies currently known. This expanded perspective may lead humans to feel insignificant and powerless in the face of such enormity. It was a big enough transition when the Copernican revolution de-centred planet Earth from the position it occupied in the Ptolemaic universe! This newer perspective could necessitate an even bigger transition.

There is a positive side to this. Abrams and Primack argue that we, humans, need to 'ground ourselves in something real that is greater than we are. Thinking cosmically can change our behaviour globally, but to think cosmically we must begin to see through cosmic metaphors. By

---

22 'Cosmology means two very different things. For anthropologists, who study human cultures, "cosmology" means a culture's Big Picture, its shared view of how human life, the natural world, and God or the gods fit together … For astronomers and physicists, the word "cosmology" means something quite different: it is the branch of astrophysics that studies the origin and nature of the universe as a whole' (Primack & Abrams, *View from the Centre*, p. 16). With Primack and Abrams I will 'seek to connect these two different understandings of cosmology by offering a science-based explanation of our human place in the universe' (ibid.).

"cosmic metaphors" we do not mean just figures of speech but *mental reframings of reality itself.*^{'23} Theirs is an appeal to think differently about the cosmos and our place in it. In language that is provocative for those seeking to move from an anthropocentric or human-centred world view to one that is anthropocosmic, Abrams and Primack proclaim that humans are at the centre of the universe, albeit a universe where Planet Earth is known to be part of the Milky Way galaxy, and that galaxy is just one in a sandstorm of other galaxies. They go on to explain that

> [t]here is no geographic centre to an expanding universe, but we are central in several ways that derive directly from physics and cosmology – for example, we are in the centre of all possible sizes in the universe, we are made of the rarest material (stardust), and we are living at the midpoint of time for both the universe and the earth.[24]

They go further to argue strongly for a '*social consensus*'[25] on how to think about the big, cosmological picture. Their understanding is that this consensus would be built on a scientific consensus that has been shared with other sectors of society to the point where together they can reflect on its significance for themselves and society.

Earth Link takes up this powerful exhortation to develop a cosmology and facilitate people's claiming the power that we have. It is a materialist, scientific cosmology that has implications for theology and spirituality. It is a further development from the heliocentric world that revolutionised the classical or traditional world view. This cosmology is understood in terms of Einstein's theory of general relativity, the modern theory of time, space and gravity. Its message about the place of humans in the universe is a salutary one, one that needs to be taken seriously if we are to move beyond 'tired (religious) metaphors from an admittedly unreal world.'[26]

Understanding evolutionary emergence

---

23 Primack & Abrams, *View from the Centre*. pp. 242–3; italics are the authors'.
24 Ibid., p. 7.
25 Ibid., p. 19.
26 Ibid., p. 269.

The story of the unfolding of the universe provides the cosmological and biological underpinnings for the connectedness of all species with all that emerged before them in that story. It is a foundational story for Earth Link. Our universe, our planet, life and humanity have a 'big history'.[27] The twentieth century afforded scientists the technological skill to detect rays coming from the origin of the universe as we know it, which is currently dated at about 13.75 billion years ago. Within the broad sweep of evolutionary history, there have been many moments when something more emerged from what was there before. Swimme and Tucker use language that is a blend of science and poetry when they present the *Journey of the Universe*.[28] They talk about the origins and unfolding of the universe as story, a new mythic yet scientifically based story to replace literal tellings of origins as occurring in seven days, or simultaneously with the beginning of human history. The scientific story, based on now measurable data, indicates that with the origins of space and time came the dynamics of inflation and gravity. In that fiery furnace, matter emerged – elementary particles combined to form the building blocks, protons and neutrons, which fused together to form the first nuclei and eventually atoms.

The story goes on to tell us how the cosmic cloud fractured and galaxies formed as gravity drew the smaller clouds together. Within these galaxies, stars are 'fiery cauldrons of transformation'[29], and one such supernova explosion gave birth to our solar system and to Planet Earth within it. Through its 4.6 billion year history, Earth, our planetary home, has gradually cooled and formed into an egg-like structure with a molten core, a crust and an atmosphere. The emergence of the first procharyotic cells about four billion years ago later led to the formation, about two billion years ago, of eucharyotic cells with their potential for the origins of life. Early forms of life progressively became more complex and have evolved into the range of species, including our own, with which we are familiar. About four million years ago, apes began the sequence of developments that gave rise to our species, *Homo*

---

27 The term 'big history' describes the history of the universe from its origins to the present.
28 Swimme & Tucker, *Journey of the Universe*. This is the source of the description that follows.
29 Ibid, p. 31.

sapiens. We humans, like all other species, have emerged from what went before us, and we are most probably not the final chapter in the story. In the words of Brian Swimme, 'This is the greatest discovery of the scientific enterprise: You take hydrogen gas, and you leave it alone, and it turns into rosebushes, giraffes, and humans.'[30]

Key evolutionary understandings

From the perspective of Earth Link, two things stand out in the story of the evolutionary emergence of organisms. Firstly, there is a dynamic within the process itself that enables the movement from simple to more complex organisms, and secondly, species have a common origin. Darwin is a major exponent of the interdependence of species based on their common origin. Elizabeth Johnson was keen to hear the message of evolutionary emergence through a study of Darwin's treatise *On the Origin of Species*. Johnson sums up Darwin's book when she says: '*On the Origin of Species* is one long argument that species are in motion, coming into being from previous species by a process that can be explained naturally, without appeal to a supernatural cause.'[31]

Darwin begins his argument by establishing the reality of variation in species, a position that challenges the understanding that they are immutable. From this starting point, it flows that the survival of a species is not guaranteed. The theory of natural selection is built on the premise that favourable variations within a species are preserved and injurious variations rejected, and the less adapted species die out over a period of time.[32] The longer-term result of these incremental adaptations is the spread of a desirable trait through the species. Ultimately, over geological periods, new species originate in dynamic relations with their surrounding communities. The result, as Johnson notes, is that organisms are 'ever more beautifully adapted to their life's situation.'[33] Darwin recognises natural selection as a 'truly wonderful fact.'[34] Johnson in turn sums up what she calls his 'profound ecological insight':

---

30 Bridle, 'Comprehensive Compassion'.
31 Johnson, *Ask the Beasts*, p. 27.
32 Ibid., p. 56.
33 Ibid., p. 56.
34 Darwin, *Annotated Origin*, p. 128.

All organic beings, living and dead, are related to one another, historically and biologically. All take their place in a single narrative of creative struggle, divergence, thriving, death, extinction, and further breakthrough. Common descent with modification by natural selection is the explanatory principle which interprets how species originate from one another, naturally.[35]

Darwin's theory was based on observation, supported by the study of fossil records throughout time and evidenced spatially in the variations of plants and animals around the globe. In a culture in which both similarity and difference were interpreted as the result of multiple separate acts of creation, Darwin drew up scenarios for the migration of species from their places of origin that established both common origins and variation brought about by adaptation to local conditions.

Darwin, in concluding, stated that his scientific observations led him to conclude that 'all organic beings have descended from one primordial form, into which life was first breathed.'[36] He acknowledged the difficulty of accepting that the dynamics of natural selection could be the workings of the natural world, rather than seeing them as independent acts of creation. While his work called for a serious appraisal of cultural and religious assumptions, his appreciation of the natural word was undimmed:

> There is grandeur in this view of life, with its several powers, having been originally breathed into a few forms or into one; and that, whilst this planet has gone cycling on according to the fixed law of gravity, from so simple a beginning endless forms most beautiful and most wonderful have been, and are being, evolved.[37]

Johnson draws her own working understanding of evolution from this work of Darwin, an understanding that is relevant for Earth Link. She extracts key features of evolution as a whole:

> *Kinship* is a most striking one: all living beings on the planet are interrelated by common descent. *The emergence of novelty* is another impressive element: new forms of life never before seen appear in the

35 Johnson, *Ask the Beasts*, p. 65.
36 Darwin, Annotated Origin, p. 484.
37 Ibid., p. 488.

course of time with new properties and abilities amid new networks of relationships. *Cumulative bodily relationship* is yet another: these new life-structures are not assembled from scratch but take shape through modifications made to already-existing simpler forms, to the point where the accumulation of many small changes leads to new organs with advanced form and function (recall the eye), and eventually even new species. *Death* is another feature of the story, a sobering companion of this biological creativity. In a *finite* universe the logic is inescapable: new patterns can only come into existence if old ones dissolve to make place for them. Seen in retrospect, a *trend toward complex organization* also characterises the process. While evolution wanders, diverges into dead-ends, indeed does not aim at any goal beyond successfully fitting an organism to its surroundings at any particular moment, its results over time show an inbuilt propensity to produce beings of ever more complicated structures by elaborating on simpler structures that already exist. Once life ignites from inorganic matter, living creatures evolve to the point of being conscious and then self-conscious, each capacity a function of increasingly organised nervous systems and brains.[38]

These key messages about the interconnectedness of all species were radical in their time and strongly influenced the transition to a new world view that can rightly be called evolutionary, the world view espoused by Earth Link.

The other important understanding about evolutionary emergence is that nature is unfolding according to a *dynamic within it.* Such an understanding renders obsolete any reliance on the direct and immediate intervention of a Creator God for nature's forward movement. Something new can come into existence from what preceded it. Emergence theory basically seeks to explain the relationship between what emerges and what went before. It offers valuable insights into emerging realities, such as mind emerging from matter.

Niels Gregersen offers the broadest view of what emergence theory is:

Emergence theory was formed as a meta-scientific interpretation of evolution in all its forms – cosmic, biological, mental and cultural

---

38 Johnson, *Ask the Beasts*, p. 102 (italics are mine).

– by British scientists in the 1920s. Although emergentists differ in their metaphysical orientation, they usually share three tenets:

- Emergents are qualitative novelties that should be distinguished from mere resultants, which come about by a quantitative addition of parts. [...]

- Nature is a nested hierarchy of ontological levels, so that the higher emergent levels include the lower levels in which they are based.

- Higher levels are not predictable from our knowledge of their constituent parts, and their operations are in principle irreducible to their lower levels.[39]

Murphy, in her introduction to the volume entitled *Evolution and Emergence*, spells out the meaning of emergence by juxtaposing it with a reductionist approach. 'If the basic idea of emergence is more or less the converse of reduction, and the core idea of reduction is that Xs are nothing more than Ys, then the core idea of emergence is that Xs are something over and above Ys.'[40]

In other words, we can see the emergence of the new either as a reassembling of what is, or we can accept that the new has causal properties that distinguish it from what gave rise to it.

Emergence theory offers a corrective to those who would reduce all of reality to its physical components. As Barbour says, it is possible to accept the theory of evolution without accepting the philosophy of materialism, which he does. He understands materialism in the following way:

> Materialism is the assertion that matter is the fundamental reality in the universe. Materialism is a form of *metaphysics* (a set of claims concerning the most general characteristics and constituents of reality). It is often accompanied by a second assertion: the scientific method is the only reliable path to knowledge. This is a form of epistemology (a set of claims concerning inquiry and the acquisition

---

39 Gregersen, 'Emergence and Complexity', p. 767.
40 Murphy, 'Introduction', p. 1.

of knowledge).[41]

He makes the point that the materialist position precludes any consideration of other forms of knowing, such as religious knowledge. Operating from a materialist premise rules out any possibility of a science–religion dialogue such as we are having right now. At the popular level, the debate rages between materialists whose evolutionary theory leads them to atheism, and those who reconcile their evolutionary theory with theism, even while it may lead them to a critical appraisal of some of its forms. The latter is the position adopted by Earth Link.

The Anthropocene

The contemporary context is being identified by many as a new geological epoch, the Anthropocene, named because of the planetary power exercised by humans. This is a time for choosing the future we desire. For some that could be *technozoic*, dominated by technological developments and fixes for many of our concerns, such as global warming. In *Laudato Si*, Pope Francis enters the conversation about whether this is the desired future. He details the symptoms of the current ecological crisis as manifest in pollution and climate change: issues around water, loss of biodiversity, changes in the quality of human life and the breakdown of society, and global inequality.[42] As noted above, he states that 'our common home is falling into serious disrepair,'[43] and I agree. Pope Francis looks beneath the symptoms of the current ecological crisis to an analysis of its causes, pointing to what he calls the dominant technocratic paradigm. He acknowledges the benefits that have come from technological knowledge but recognises that its basic premise is flawed. 'This paradigm,' he says, 'exalts the concept of a subject who, using logical and rational procedures, progressively approaches and gains control over an external object.'[44] He goes on to say that

> [t]he specialization which belongs to technology makes it difficult

---

41 Barbour, *Nature, Human Nature*, p. 4.
42 *Laudato Si,* pp. 20–60.
43 Ibid., p. 61.
44 Ibid., p. 106.

to see the larger picture. The fragmentation of knowledge proves helpful for concrete applications, but it leads to a gradual loss of appreciation for the whole, for the relationships between things, and for the broader horizon, which then becomes irrelevant. This very fact makes it hard to find adequate ways of solving the more complex problems of today's world, particularly those regarding the environment and the poor; these problems cannot be dealt with from a single perspective or from a single set of interests.[45]

The preferred future of Pope Francis is one of integral progress, one in which social, economic and environmental concerns are addressed together.

One aspect of contemporary society that comes under scrutiny in *Laudato Si* is 'misguided anthropocentrism'[46], which ignores the fact that 'everything is interrelated,'[47] a view that is developed at length in this book. Currently, in the Anthropocene, humans are present to this planet in such a way as to be a planetary power. A group of international scientists have explained what this means:

Over the past century, the total material wealth of humanity has been enhanced. However, in the twenty first century, we face scarcity in critical resources, the degradation of ecosystem services, and the erosion of the planet's capability to absorb our wastes. Equity issues remain stubbornly difficult to solve. This situation is novel in its speed, its global scale and its threat to the resilience of the Earth System. The advent of the Anthropocene, the time interval in which human activities now rival global geophysical processes, suggests that we need to fundamentally alter our relationship with the planet we inhabit. Many approaches could be adopted, ranging from geoengineering solutions that purposefully manipulate parts of the Earth System to becoming active stewards of our own life support system. The Anthropocene is a reminder that the Holocene, during which complex human societies have developed, has been a stable, accommodating environment and is the only state of the Earth System that we know for sure can

45 Ibid., p. 110.
46 Ibid., pp. 115–22.
47 Ibid., p. 120.

support contemporary society. The need to achieve effective planetary stewardship is urgent. As we go further into the Anthropocene, we risk driving the Earth System onto a trajectory toward more hostile states from which we cannot easily return.[48]

Figure 2 is a diagram from the Swedish Academy of Sciences indicating the safe operating space for nine planetary systems. These have already been exceeded for the rate of biodiversity loss, climate change, and human interference with the nitrogen cycle, while the boundaries for the phosphorus cycle, ocean acidification, global freshwater usage, and changes in land usage are moving towards the limit of the safe operating level.

The relationships between humans and the planet are in a critical transition, during which we need to acknowledge and redress our negative impacts on the planet. If anything, the vision and mission of Earth Link gain greater urgency from these developments, even as Earth Link takes heart in the public profile that environmental degradation and human poverty are now being given.

The messages from the sciences are clear: there is an unfolding dynamism in the universe and in the evolutionary emergence of species; everything is interconnected and interdependent, even to what can be considered kinship relations between the species; and within the interconnectedness, entities retain their distinctive characteristics. These messages are embraced by Earth Link. They are foundational in the development of ecospiritual understandings and practices, and their theological underpinnings. Principle 2 is about Earth Link's approach to ecospiritual understandings and practices, and a beginning response to the urgency of the global environmental situation.

---

48 Steffen et al., 'Anthropocene', pp. 739–61.

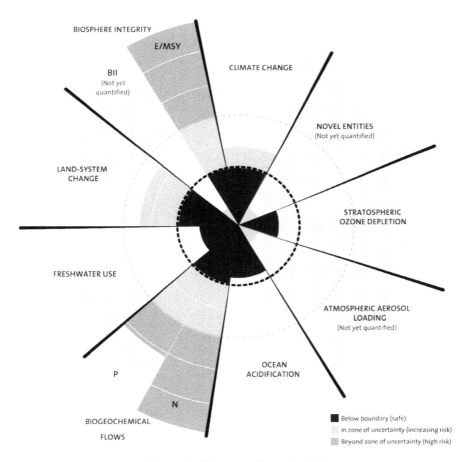

Figure 2: Planetary Boundaries[49]

---

49 Steffen et al., 'Planetary Boundaries'. Used by permission.

CHAPTER 5

Principle 2: Mystical Encounters

*Cultivate an open, attentive and receptive attitude
in order to enter into transformative, mystical encounters.*

This Principle captures Earth Link's conviction that the ecospiritual experience is about transformative, mystical encounter, and that it can be facilitated, but not guaranteed, by an open, attentive, receptive attitude. Given the influence of Thomas Berry on Earth Link from its inception, this enhanced principle is not startlingly new but rather builds upon and restructures original material while adding further insights. In this enhanced principle I will consider ecospirituality, both subjectively and objectively, before looking at ways of facilitating encounter. Mystical encounter is understood or interpreted according to individual or communal perspectives. This first principle draws on elements of a Judaeo-Christian approach as sources of images of God that can be meaningful for ecospirituality. Subsequent principles draw on Catholic ecotheology as a way of interpreting and deepening the ecospiritual experience.

What is ecospirituality?

Ecospirituality is about transformative, mystical encounter. Earth Link has been profoundly influenced by Thomas Berry, not only in his drawing attention to the planetary crisis and his diagnosis of the root cause as the disconnection of humans from nature, but also by his personal testimony to the spirituality that built on his transformative experience of the meadow. In *The Great Work,* he referred to an experience in his childhood that became normative for him:

It was an early afternoon in May when I first wandered down the incline, crossed the creek, and looked out over the scene. The field was covered with white lilies rising above the thick grass. A magic moment, this experience gave to my life something that seems to explain my life at a … profound level. It was not only the lilies. It was the singing of the crickets and the woodlands in the distance and the clouds in a clear sky. […]

This early experience … has become normative for me throughout the entire range of my thinking. Whatever preserves and enhances this meadow in the natural cycles of its transformation is good, what is opposed to this meadow or negates it is not good. […]

That is good in economics which fosters the natural processes of this meadow. So in jurisprudence, law, and political affairs—what is good recognizes the rights of this meadow and the creek and the woodlands to exist and flourish in the ever-renewing seasonal expression. […]

Religion too, it seems to me, takes its origin here in the deep mystery of this setting. The more a person thinks of the infinite number of interrelated activities taking place here, the more mysterious it all becomes.

The more meaning a person finds in the Maytime blooming of the lilies, the more awestruck a person might be in simply looking out over this little patch of meadowland.[1]

Berry's ecospirituality builds on this 'magic' experience, out of which he also makes meaning about what is good and what needs to be addressed in the major systems of society, be they religious, social, economic, legal, political, or environmental.

A transformative experience of the meadow formed the basis on which he built his spirituality, and this enabled him to call others to moments 'akin to ecstasy'.[2] This is an intimate experience of the natural world as subject rather than as an object to be exploited.[3] His spirituality is even more: it is recognition of the numinous qualities of Earth.[4] It is

1 Berry, *Great Work*, pp. 12–13.
2 Berry, *Sacred Universe*, p. 132.
3 Berry & Swimme, *Universe Story*, p. 243.
4 Berry, *Sacred Universe*, p. 73.

the basis for his sense of the divine, a sense that is diminished with the destruction of ecosystems. Berry elevated the unfolding story of the universe to the status of a mythic, sacred story, and acknowledged the book of nature as the primordial source of God's revelation. This was the grounding of his ecospirituality, and it has taught Earth Link to value immersion in nature as key to the 'deep bonding' that is at the heart of its mission.

Berry's spirituality, shaped as it is in an emerging universe, is not only a Gaian sense of Earth as a living system, but also 'a mode of being in which not only the divine and the human commune with each other but through which we discover ourselves in the universe and the universe discovers itself in us.'[5] In this he was influenced by the writings of Teilhard de Chardin, who had a deep sense of divine presence within the very structures of the universe. Heaven and Earth embrace in the emergent, evolutionary spirituality of Thomas Berry.

Bron Taylor, in his study of nature religion, understands spirituality as 'a quest to deepen, renew, or tap into the most profound insights of traditional religions, as well as a word that consecrates otherwise secular endeavours such as psychotherapy, political and environmental activism, and one's lifestyle and vocational choices.'[6] Schneiders, in her academic study of spirituality, sees it as 'the actualization of the basic human capacity for transcendence' and 'the experience of conscious involvement in the project of life-integration through self-transcendence toward the horizon of ultimate value one perceives'.[7] While spirituality is of its nature a human activity, Schneiders is holistic in her attitude to persons and recognises that life is lived within a context that influences one's beliefs, attitudes and behaviours. Contemporary anthropological spirituality is 'not only a complexification of the holistic approach to the human subject of religious experience, but a heightened awareness of the dimensions and influence of "place" and "space" (both inner and outer), globalization, ecological crises, the validity of religious experience outside one's own tradition, scientific developments, and

5 Ibid., p. 72.
6 Taylor, *Dark Green Religion*, p. 3.
7 Schneiders, 'Christian Spirituality', p. 2.

cultural currents.'[8] Ecospirituality is an expression of this heightened awareness, and is the focus of the resourcing, reflecting and acting of the Earth Link community.

Ecospirituality involves emotion. It is experiential. The emotions that arise during one's immersions in nature are multilayered. One can feel pain when listening to the cry of Earth and her people, and rage, frustration and disillusionment at policies and decisions that jeopardise their health and wellbeing. One can be filled with awe and wonder on encountering beauty, sensing the enormity of the galaxies and marvelling at the unfolding story of our universe. One can experience a deep sense of connectedness to place and marvel at the complexity evident in Darwin's 'tangled bank', and at the processes of natural selection that brought it to this point in time. One can experience pain and sorrow on encountering natural disaster, tragedy or loss, or suffer from the inevitable limits of our finite existence and the effects of evil. Yet in the midst of this, one can be suffused with hope, with a deep sense of Earth as God's dwelling place and encounter what is hidden even while it is being revealed. And one can struggle to make meaning of one's experiences with an outmoded worldview, spirituality and theology that are not cognisant of developments from cosmology and from natural and behavioural sciences.

Awe and wonder are natural emotions that are often, but not exclusively, associated with religious experience. In the words of Fuller, writing on the connection between wonder and spirituality, wonder is 'an emotional response that promotes passive, receptive modes of attention in the presence of something unexpected rather than fight or flight responses.'[9] Wonder arises spontaneously and creates openness. Fuller continues as follows:

> First, wonder is an emotion linked with approach and affiliation rather than avoidance. Wonder motivates attention and motivates a quest for increased connection and belongingness with the putative source of unexpected displays of life, beauty, or truth. Wonder is thus somewhat rare among the emotions in its functional capacity

8 Ibid., p. 10.
9 Fuller, 'Wonder and Spirituality', pp. 364–84.

to motivate persons to venture outward into increased rapport with the environment. Second, wonder awakens our mental capacity for abstract, higher-order thought. Indeed, wonder seems to direct our cognitive activities to identify causality, agency, and purpose in ways that are not directly connected with our biological survival.[10]

In another context, Swimme and Tucker say that wonder will guide us. For them, and hopefully for us, 'wonder is a gateway through which the universe floods in and takes up residence within us.'[11] Contemporary ecospirituality needs to be cosmic in scope and grounded in awareness of the dynamic unfolding of the universe. In the tradition of Berry, Earth Link has adopted this story as foundational. The telling and re-telling of the universe story locate us within a much greater whole, enabling us to have a cosmic identity as well as a planetary one. This story reminds us to cultivate 'a different kind of "*spirituality-of-being in-the-world*", a *spirituality of being connected* to the ordinary life in the world with its daily relationships and responsibilities, a spirituality that makes sense of our environment without and within.'[12] This is a spirituality that ensures we are familiar with the dynamics of our finite existence as we remember that 'the finite bears the infinite.'[13] Larry Rasmussen, in *Earth Community, Earth Ethics*, develops this motif when he says, 'God is pegged to earth. So if you would experience God, you must fall in love with earth. The infinite and transcendent are dimensions of what is intensely at hand. Don't look "up" for God, look around. The finite is all there is, because all *that* is, is *there*.' [14]

Facilitating encounter

While spiritual encounters can be unmediated, Earth Link is committed to facilitating the development of an open, attentive and receptive attitude in order to enter into transformative, mystical encounters. Earth Link promotes practices that involve immersion in sacred place, formation of a sacramental mindset toward nature, and individual and

---

10 Ibid., p. 370.

11 Swimme & Tucker, *Journey of the Universe*, p. 113.

12 King, 'One Planet, One Spirit', pp. 66–87, 72 (italics by the author).

13 Rasmussen, *Earth Community*, p. 273.

14 Ibid.

communal ritual practices. To a lesser extent Earth Link promotes the cultivation of virtue as part of its ecospirituality. Earth Link would agree with Gottlieb:

> (Thus) nature spirituality in the modern age contains within itself the seeds not only of a personal but also of a social transformation. If the personal ego can make us sick as individuals, our collective ego of reckless industrialization and unrestrained consumption can make us ecologically sick as a society. If yoga and meditation are helpful responses to individual maladies of modernity such as high blood pressure and anxiety, so a spiritually oriented relation to nature may be our best response to the collective maladies of pollution and climate change. [15]

I will elaborate on some ways of facilitating encounter with Sacred Mystery.

Immersion in sacred place

Falling in love with Earth as valuable in itself, and as a medium of divine revelation, requires the knowledge and intimacy that comes from immersion. Familiarity with place – indeed, developing a physical relationship with it – is foundational to an openness to transformative, mystical encounters. As acknowledged above, such encounters can be unmediated, yet immediacy of contact with Earth and the cosmos is an important aspect of ecospirituality. In the preface to *Landscapes of the Sacred,* Schneiders says, 'In our experience of the sacred the "where" is as determinative as the "how".' [16] Earth Link captured this insight in the first edition of the Principles by noting that 'Earth Link spirituality is real, immediate and tangible; a spirituality in which the human body is called to be in dynamic contact with the Earth, "profoundly grounded in the realities of the earth and the body."' [17]

This 'where' is cosmic as well as planetary. Berry strongly advocated developing a cosmic identity. For him, there was a mutual relationship between physical bodies and the universe. In an oft-quoted verse he says:

---

15 Gottlieb, *Spirituality*. See Chapter 9, 'Spirituality and Nature', p. 3.
16 Lane, *Landscapes*, p. iv.
17 The quote within the quote is from Tacey, 'Spirit Place'.

The child awakens to a universe.
The mind of the child to a world of wonder.
Imagination to a world of beauty.
Emotions to a world of intimacy.
It takes a universe to make a child both
in outer form and inner spirit.
It takes a universe to educate a child.
A universe to fulfill a child.
Each generation presides over the meeting of these two
in the succeeding generation.
So that the universe is fulfilled in the child,
and the child is fulfilled in the Universe.
  While the stars ring out in the heavens![18]

Belden Lane cogently makes the point that religious experience is invariably 'placed' experience and that those places are frequently the most ordinary ones entered anew with awe.[19] Lane posits four axioms to assist us in our understanding of place as sacred, which I offer in summary:

- Sacred place is not chosen, it chooses.
- Sacred place is an ordinary place made extraordinary through ritual.
- Sacred place can be trodden without being entered.
- The impulse of sacred place is both centripetal and centrifugal, local and universal – at times the quest is for centredness, and at other driven out from that locale.[20]

Immersion in place is of particular importance to indigenous peoples. In Australia, the first peoples have a continuing sense of Earth as sacred. Indigenous poet, Denis Kevans, rebukes those for whom the land is just a quarry to be mined, as he reminds us:

---

18 Thomas Berry, *It Takes a Universe,* http://thomasberry.org/publications-and-media/it-takes-a-universe.
19 Lane, *Landscapes*, p. 28.
20 Ibid, p. 15.

Sacred means that ... sacred ... it's a place where spirits rise,
with the rainbow wings of sunset, on the edge of paradise,
Sacred, that's my father, that's my mother, that's my son,
Sacred ... where the dreaming whispers hope for everyone.[21]

A sense of the sacredness of Earth has been found to be a motivating force in environmentally responsible behaviour. Hedlund-de Witt[22] interviewed nature-lovers/environmentalists and spiritual practitioners in Victoria, Canada, who were already behaving in an environmentally responsible manner. She identified three distinct pathways to environmental responsibility: profound encounters with nature, contemporary spirituality, and some blend of the two. Taylor abstracted from many case studies his inclusive notion of nature religion as 'an umbrella term to mean religious perceptions and practices that are characterised by a reverence for nature and that consider its destruction a desecrating act. Adherents often describe feelings of belonging and connection to the earth — of being bound to and dependent upon the earth's living systems.'[23] Ecospirituality within an emerging universe has the potential to facilitate a deeper sense of the embrace of Heaven and Earth, and contribute to redressing environmental degradation.

Formation of a sacramental mindset

The sense of connection that can come from an open, attentive and receptive attitude can be understood in terms of a relationship with 'animate earth'[24] in the Gaian sense of Earth's having the qualities of a living organism. For others such as Berry and Lane, the experience of Earth's inner numinous quality, or the experience of Earth as sacred, can be an encounter with divine Mystery. It is sacramental in the broadest sense of that word. For Lane, this is about having a binary vision that

21 Denis Kevans, in *Ah, White Man, Have You Any Sacred Sites?*, 1985 [ISBNO 9593 073], quoted at: http://www.nyungah.org.au/documents/AhWhiteM.html. Kevans wrote the poem 'as a reply to Hugh Morgan (Western Mining Company), [who] accused Aboriginals of being 'pagans' for opposing the Noonkanbah mining project in 1984' (ibid.). Reproduced with permission.
22 Hedlund-de Witt, 'Pathways to Environmental Responsibility ', pp. 154–86.
23 Taylor, *Dark Green Religion*, p. 5.
24 Harding, *Animate Earth*.

can hold the ambivalence of seeing and valuing the ordinary, while recognising that 'the ordinary is no longer at all what it appears.'[25] There are different understandings of the nature of the encounter. Johnson, in her Trinitarian theology, refers to an encounter with God's presence based on the understanding that God is in all things and places but not contained by them. Contemporary theology calls this relationship *panentheism*, which to Johnson means that 'the world is indwelt by the presence of the Spirit while at the same time it is encompassed by divine presence which is always and everywhere greater.'[26] Johnson distinguishes panentheism from pantheism, which 'conflates God and the world,'[27] and from unipersonal theism, 'which posits God as a transcendent cause.'[28] Johnson espouses this widely held position of panentheism as respecting both the immanence and transcendence of God, even as it opens up 'a kind of asymmetrical mutual indwelling, not of two equal partners, but of the infinite God who dwells within all things sparking them into being and finite creatures who dwell within the embrace of divine love,'[29] insights that are of vital importance to the practice of ecospirituality within an emerging universe. This is when Heaven and Earth embrace without obscuring the differences, a position endorsed by Earth Link. Principles 3 to 5 explore how the Christian tradition understands Divine Mystery, with whom one enters into a mutual but not equal relationship.

Ritual, both personal and communal, is very important in the creation of an intentional space for entering into transformative mystical encounters. Earth Link has always engaged in ritual, along with holistic processes, in order to awaken people to spiritual insights and encounters. Tucker and Grim recognise that religions at their best have 'woven humans into nature with rituals, symbols and ethical practices.'[30] It is our hope that religious and spiritual practices also weave humans into the presence of the Spirit embedded in nature yet

---

25 Lane, *Landscapes*, p. 37.
26 Johnson, *Ask the Beasts*, p. 147.
27 Ibid.
28 Ibid.
29 Ibid.
30 Grim & Tucker, *Ecology and Religion*, p. 11.

not contained by it. Ecospiritual practices have the potential to facilitate a truly sacramental orientation to matter. Many approaches are used within ecospiritual practice. Some have the potential to awaken people to the richness of existing ecclesial practices, many of which had their origins when people had a greater sensitivity to the rhythms and cycles of nature. Ecospiritual practices celebrate the hours of the day, the cycles of sun and moon, and the seasons. Time-honoured contemplative practices such as *Lectio Divina* can be broadened to include listening to the book of nature as well as the book of scripture. Earth Link currently espouses many of these practices and adapts them to the circumstance of the particular ritual and group.

Encountering the Sacred in wind, water, fire, bird and wisdom

Ritual practice has been built since early times on a sense that 'the finite contains the infinite,' even while acknowledging that the infinite is not contained by the finite. The elements of water, wind, fire and earth form the basis of the sacramental system, yet the sacramental mindset referred to above is broader than just the seven sacraments. While later principles elaborate on the doctrine of continuous creation and of God's indwelling presence, this second principle is enriched by Johnson's retrieval of cosmic and planetary biblical images of God, and their potential for enhancing our understanding of the Judaeo-Christian tradition's understanding of the Sacred. This biblical material contributes to our naming whoever is revealed in our encounters with cosmos and planet.

In *Ask the Beasts,* Johnson unpacks a range of cosmic images that expand 'the notion of divine presence beyond analogy with a human person.'[31] She retrieves the cosmic, biblical images of blowing wind, flowing water and blazing fire, along with that of the bird, before tapping into the Wisdom tradition. These references can open our eyes to a long tradition of using analogies drawn from nature to capture a sense of God's presence, and, conversely, the encounter with natural forces can be interpreted with reference to the presence of Holy

---

31 Johnson, *Ask the Beasts,* p. 134.

Mystery. It is with reference to the mutuality of these processes that this principle engages with these images of divine Mystery that can be part of the ecospiritual journey.

Images of God as the male, authoritarian figure in the sky or in heaven do not serve us well as we seek to deepen our appreciation of God's continuous presence in the community of life. The biblical tradition is rich with cosmic images that have been ignored for as long as humans have been at the centre of our world view and theology. For Johnson these images expand 'the notion of divine presence beyond analogy with a human person.'[32] They also convey the dynamism inherent to images of powerful natural forces such as blowing wind, flowing water and blazing fire. In Johnson's words, '[T]hey can surround and pervade other things without losing their own character; their presence is known by the changes they bring about.'[33] The animal metaphor that Johnson uses, the bird, long associated with the sacred in primal religions, is an image that conjures up the experience of brooding and flying free. Her image of Holy Wisdom is an extension of her earlier Trinitarian work, this time with attention to Wisdom's cosmic reach. For Johnson and for us, these images open up new avenues for thinking about God's presence in the natural world, and for providing language for articulating such insights. This opening up of ways of imaging God is of particular relevance for ecospirituality. The biblical references can open our eyes to a long tradition of using analogies drawn from nature to capture a sense of God's presence, and, conversely, the encounter with natural forces can be interpreted with reference to the presence of Holy Mystery.

Johnson is quite lavish in her exploration of these images. She piles them up one on top of the other and accumulates them into a rich picture of the dynamic presence of God in what she refers to as the community of life, which can also be referred to as the Earth community. She begins with '*ruach*', understood as wind or breath. When Earth Link was located on an escarpment of the D'Aguilar Range, we called the cottage by that name. It had the biblical connotation but also reflected

---

32 Ibid.
33 Ibid.

the weather at that site! One of the things you notice about wind is that you cannot see it unless it is carrying particles, yet you can definitely feel it. Breath is a vital sign of life. In the biblical world of Genesis 1:1–2, the *ruach* of God breathed the world into existence. It breathed life into dry bones and reminded the people that the same could happen to them (Ezek 37:1–4). The wind filled the room where the frightened disciples were gathered in the wake of the death of Jesus and filled them with courage in the midst of their confusion (Acts 2:2–4).

Johnson moves from the dynamism of wind to various forms of water, whether trickling or rushing, fresh or salt. Water makes up the biggest part of living organisms. To the people in exile wandering in the desert, Isaiah spoke of God's faithful presence in terms that they would understand: 'I will pour water on the thirsty land, and streams on the dry ground' (Is 44:3), and he assured them that God would bless their descendants. Again, when the disciples were crowded together in fear, Peter reminded them of Joel's promise that God would pour out his Spirit on them (Acts 2:18). Augustine made his own contribution to the understanding of God's presence permeating all of creation by likening it to a sponge saturated in water.[34] God's presence is not stinted but superabundant.

The picture builds up with the image of fire as a symbol of God's fiery and enlightening presence. Like wind and water, fire can also be destructive and symbolic of God's wrath. The book of Exodus tells us that it was in the context of a burning bush that Moses received the revelation of who God is and of his enduring presence with them on their journeys. God's flaming presence in the tongues of Pentecost fire set their hearts ablaze (Acts 2:3-4). Even Stephen Hawking, whose atheism was well known, and whose knowledge of and marvel at the wonders of the universe were also well known, asked the question, 'What is it that breathes fire into the equations and makes a universe for them to describe?'[35] For those with a faith orientation, that fire is an analogy for the spark of the life-giving presence of the Spirit.

In the Mediterranean region, the bird is an ancient symbol of the

---

34 Ibid., p. 137.
35 Hawking, *A Brief History of Time*, p. 174.

feminine deity, and in contemporary times it is the widely used symbol of the peace dove. Crossing over to the Hebrew religious tradition, the protective aspect of the bird's wings features in passages such as Psalm 17:8: 'Guard me as the apple of the eye; hide me in the shadow of your wings.' Early Syriac Christian imagery uses the brooding, motherly bird to describe the actions of the Spirit towards her children, and it is present in the conception, birth and death of Jesus. Augustine comments on the Genesis creation text in which 'the Spirit of God swept over the face of the waters' (Gen 1:2), likening that image to one in which 'the warmth of the mother's body in some way also supports the forming of chicks through a kind of influence of her own kind of love.'[36] This animal image draws attention to the creative, nurturing role of the Spirit and enhances our understanding of the Spirit's relational presence.

In her final image Johnson revisits the Wisdom tradition, which featured prominently in her feminist imaging of God-Sophia, Jesus-Sophia and Spirit-Sophia. She again quotes Augustine, who speaks of how the Wisdom of God 'took our weakness upon herself and came to gather the children of Jerusalem under her wings as a hen gathers her chicks.'[37] This is multilayered Wisdom Christology. Johnson points to the way in which the biblical figure of holy Wisdom has cosmic scope. She is present with God at creation (Ps 8), is a teacher of 'what is secret and what is manifest' (Wis 7:22), is present, 'pervading and penetrating all things' (Wis 7:24), and defeats all evil (Wis 7:29–30). This image of Wisdom points to the all-pervading presence who 'reaches mightily from one end of the earth to the other, and (she) orders all things well' (Wis 8:1). Wisdom is identified with God's Spirit in her attributes (Wis 7:22), and, even more importantly, linked to the mystery of God's being as a 'breath of the power of God' and an 'image of his goodness' (Wis 7:25–26). Johnson points out that while there is no direct identification of God and Wisdom, the way in which she moves and functions helps our understanding of the Spirit.

Johnson concludes by reminding us that 'these symbols provide guidance for how to think about the hidden presence and activity of the

---

36 Augustine, *Literal Meaning*, Book I, p. 36.
37 Ibid., p. 26.

Spirit of God in the natural world,'[38] a presence that loves, pervades and vivifies, while remaining transcendent, incomprehensible Mystery.

The hidden presence and activity of the Spirit can be encountered through the ecospiritual experience. The felt embrace of Heaven and Earth can be not only a transformative, mystical encounter as encouraged in this principle, but also a recognition of the interconnectedness of the dynamic entity that is the universe called for in Principle 1. These principles can contribute to making real Earth Link's vision of 'a world where there is respect, reverence and care for the Earth community.'

---

38 Johnson, *Ask the Beasts,* p. 143.

CHAPTER 6

Principle 3: Indwelling Presence

*Acknowledge the indwelling presence of the Spirit
in Earth and cosmos.*

The first of the enhanced principles promotes a world view that is cognisant of scientific and cultural changes. People are invited to 'recognise that the universe is a dynamic entity, within which all is interconnected yet distinct.' The second enhanced principle relates to ecospiritual understandings and practices, and invites people to 'cultivate an open, attentive and receptive attitude in order to enter into transformative, mystical encounters.' For that principle we draw on Johnson's retrieval of cosmic and planetary biblical images of God, which expand the notion of the Divine in a way that is meaningful for ecospirituality. The enhanced principles 3, 4 and 5 develop Christian religious perspectives more overtly. They are strongly influenced by Johnson's Trinitarian theology as it pertains to the beliefs that give depth to the ecospiritual encounter. These beliefs are developed more fully in the enhanced principles 3–5 than in the earlier Earth Link material.

Earth Link has a mission to facilitate deep bonding within the community of creation. In our earlier statement of principles we expressed the conviction that such bonding was an acknowledgment of the presence of the Sacred in the interdependent web of life.[1] We believe that this is 'an experience of the Sacred, and that Earth and cosmos constitute for us a primary revelation of Ultimate Mystery.'[2] Berry goes so far as to declare this the primordial revelation in the sense that it is antecedent to revelation from scripture or from tradition. Pope Francis and others speak of revelation through the book of nature as well as the

---

1 Original Principle 3.
2 Costigan, Rose and Tinney, *Introduction to Ecospirituality,* p. 2.

book of scripture.[3] It is imperative to note that revelation is from God, omnipresent in the universe, and, while we can facilitate receptivity, we are ultimately in the hands of Divine Mystery. That Earth and cosmos are revelatory is a key concept in ecospirituality that is understood as transformative, mystical encounter. Such is the ecospirituality espoused by Earth Link.

The doctrine of revelation is a classical Christian doctrine, which Johnson revisits with a focus on the theological significance of the natural world, and with special reference to the questions raised by its evolutionary emergence. This enhanced principle of Earth Link provides a specifically Christian interpretation of the widely experienced sense of Earth as sacred. It adds the dimension of God's indwelling, which needs to be understood in ways that respect the dynamic and autonomous processes of the universe and Earth. The exposition of this principle draws extensively on Johnson, who takes as her starting point 'the presence of the Spirit throughout the world in the act of continuous creation'[4] and proceeds from there to theologise about God's transcendent presence, God's loving action in our created reality, and the relations between divine and created agency. Other aspects of Johnson's Trinitarian theology are pertinent to Principles 4 and 5.

In this context, the symbol of the Celtic triquetra, which is at the head of this chapter, can be used as a symbol of the interrelatedness of God's Trinitarian life within itself, and as encountered in cosmos and planet Earth. The elements of the triquetra – God, humans and Earth – are interconnected without losing their distinctiveness. Together they form a whole that is greater than the sum of the parts.

God's indwelling presence in continuous creation

As noted in Principle 1, we are in the throes of moving away from classical understandings of the world as static, hierarchical, and marked by dualistic understandings of the relationship between spirit and matter, body and soul, and even mind and matter. These influences made their

---

3 Pope Francis, *Laudato Si.*
4 Johnson, *Ask the Beasts,* p. 128.

way into theologies that ranked the supernatural over the natural, and saw grace, understood as God's gift[5], as superior to and separate from nature. Theologically, interest in creation waned in favour of interest in redemption, and God's action in the biblical history of salvation was seen as more significant than God's action in the cycles of nature. In terms of spirituality, the journey was one of ascent from the lesser to the greater, in a way that devalued Earth in the pursuit of Heaven. Earth Link does not subscribe to this view, but rather, with Johnson, seeks a different starting point, that of 'the presence of the Spirit throughout the world in the act of continuous creation.'[6]

Classical creation theology encompasses 'original creation in the beginning, continuous creation in the present here and now, and new creation at the redeemed end time.'[7] In the Hebraic-Christian tradition, plants and animals have a reference point beyond themselves in the 'overflowing generosity of the incomprehensible God who freely shares life with the world.'[8] God's presence is there as origin, source and goal. It is an ongoing and dynamic presence, and Johnson borrows the metaphor of the singer keeping the song in existence at all times to give the sense of immediate and sustained presence.[9] God's presence is traditionally known as the Spirit, who also 'continues to draw the world to an unpredictable future, pervaded by a radical promise at the ultimate end of time.'[10]

Continuous creation, as understood by Johnson, affirms that 'rather than retiring after bringing the world into existence at some original instant, the Creator keeps on sustaining the world in its being and becoming at every moment.'[11] Johnson finds this truth implied in the Nicene Creed with its affirmations of God as Creator in the beginning, and of Spirit as 'Lord and Giver of life.' The Spirit is the vivifier, the life-giver.

---

5 Ibid., p. 126.
6 Ibid., p. 128.
7 Ibid., p. 123.
8 Ibid., p. 124.
9 McCabe, *God, Christ and Us,* p. 103.
10 Johnson, *Ask the Beasts,* p. 124.
11 Ibid., p. 128.

Johnson searches the sources of revelation for insights that have been overlooked. The biblical tradition is replete with images of a dynamic and omnipresent God, and it is there she begins her search before turning to Aquinas for insights into the relationship between God and world.

From the Hebrew scriptures, Johnson draws attention to the life-giving action of the Spirit in moving over the waters in the process of creation (Gen 1:2), renewing the face of the Earth (Ps 104:30), filling creation with God's presence (Wis 1:7), and being intimately present everywhere. In a related verse we read:

> Where can I go from your spirit? Or where can I flee from your presence?
> If I ascend to heaven, you are there; if I make my bed in Sheol, you are there;
> if I take the wings of the morning and settle at the farthest limits of the sea,
> even then your hand shall lead me, and your right hand shall hold me fast (Ps 139:7–10).

She also points to the balance that the biblical writers find between God's transcendence and immanence. This is the God who is over and above all, but who is also present and engaged with all that is. God who holds all in the palm of God's hand is loving and compassionate. God is personally engaged in this anthropocosmic reality at all times and in all places, not only in human experience, but with deep love for all the living:

> For you love all things that exist …
> You spare all things, for they are yours, O Lord,
> you who love the living
> For your imperishable spirit is in all things (Wis 11:24; 12:1).[12]

The whole of the natural world is imbued with Spirit and is not to be denigrated or demeaned for its materiality. Heaven and Earth really do embrace.

Johnson moves to the emergence in early Christian times of the

---

12 This is the translation used by Johnson.

Trinitarian instinct about God as both ineffable and present. In the depths of their experience, the early Christians encountered Jesus present among them personally and had a sense of his continuing resurrected presence. The post-resurrection faith of the community issues forth in this blessing: 'The grace of our Lord Jesus Christ, the love of God, and the fellowship of the Holy Spirit be with you (2 Cor 13:10).'

In the second century, Irenaeus was the first to attempt to capture this Trinitarian instinct more systematically. Around that time, Tertullian in Africa came up with various metaphors for the relationship between the Trinity and nature, a reference that recurs in Johnson's work. He uses nature metaphors such as sun and water, and the root, shoot and fruit of a tree, to capture the presence of the incomprehensible God who is 'unleashed' in the world. To take just one metaphor as Johnson presents it:

> If God the Father can be likened to the sun, source of light and heat, then Christ is the ray of sunlight streaming to earth (Christ the sunbeam, of the same nature as the sun), and the Spirit is the suntan or sunburn, the spot of warmth where the sun actually arrives and has an effect.[13]

Subsequent developments in Trinitarian theology have at times become more conceptual and abstract, while at other times they are enriched with understandings of the relationality within God's being and beyond. As part of her reclaiming of Spirit, Johnson appeals to Aquinas' sense of Spirit as 'God proceeding by love,' the bond of mutual unity and love in the inner life of the Trinity and in action in the world. Earth Link acknowledges that the Spirit is indeed the 'holy mystery of God's own personal being,'[14] who is continuously present and active in the process of continuous creation, a presence that demands our conscious attention as we respond to nature with wonder and face the consequences of our wasteful treatment of it.

---

13 Ibid., p. 131.
14 Ibid., p. 133.

God's transcendent presence

In the Christian tradition, and in these principles, God's presence is not identical with natural reality, even when intimately present to it. Johnson begins her philosophical exploration of the personal presence of God for all creatures with the insights of Aquinas. In his quest to establish that 'God alone is the source of everything,'[15] Aquinas worked with the notion of God as Being itself, while all else participates in or receives that being as gift. Kasper captures this relationship as 'self-communicating love.'[16] To be in relation is not an optional extra but is of the very essence of Godself. In a Trinitarian theology this means that God's inner being can be understood as 'self-giving love beyond imagining.'[17] Without a personal conviction about this, any Christian ecospirituality is but an academic exercise.

Johnson goes on to consider the relations between God and all that is not God, a key consideration for ecospirituality. The very act of creation sets up a relationship of dependence on the source and sustainer of all. For all things to have being, they must be in relation to Being itself, and Aquinas concludes that 'God is in all things, and innermostly.'[18] Yet the distinction between Being itself and the recipient of being maintains the distinctiveness of each according to its own nature.

When we claim that we can encounter God present in the Earth community, we need, with Johnson, to ask if God is in all things. Our questions can be enlightened by her examination of Thomas Aquinas. She draws on his image of fire as she explains that

> [j]ust as fire ignites things and sets them on fire, the Spirit of God ignites the world into being. This obviously happens in the beginning but doesn't stop; just as the sun brightens the air all the day long, the presence of the Spirit sustains creatures with the radiance of being as long as these exist.[19]

Johnson uses this symbol as an analogy for the Spirit dwelling in the

---

15 Ibid., p. 144.
16 Kasper, *God of Jesus Christ*, p. 156.
17 Johnson, *Ask the Beasts,* p. 145.
18 Aquinas, *Summa Theologicae*, I,8,1.
19 Johnson, *Ask the Beasts*, p. 146.

innermost regions of the whole universe. The omnipresence of the Spirit is not tangible. It does not occupy space. Speaking of God's presence is faith language for the understanding that God is in all things and places but not contained by them. Aquinas also argues that all things are in God. Contemporary theology calls this relationship panentheism, and I refer to it in Principle 2. It is a widely held position, seen as respecting both the immanence and transcendence of God, even as it opens up 'a kind of asymmetrical mutual indwelling, not of two equal partners, but of the infinite God who dwells within all things sparking them into being and finite creatures who dwell within the embrace of divine love.'[20] These are insights which are of vital importance to Earth Link's practice of ecospirituality within an emerging universe. This is when Heaven and Earth embrace, yet in a way that respects their differences.

Theologians use the notion of participation to further understanding the relationship between God and all that is not God. Norris Clarke explains participation as having three elements: an infinite source, finite things and a link between the two.[21] Johnson explains Aquinas as saying that God in creating bestows vitality 'in a creaturely way to what is other than Godself.'[22] In accepting this gift of being, all creatures share or participate in Being itself. The gift is freely given, and the receiver relies on the gift. This participation applies to all of the natural world, 'not divinely, but as created, that is, existing and acting according to its own finite nature.'[23] Johnson gives an example by considering 'goodness.' God alone is good, but creation, which is rich and exuberantly alive, partakes in that goodness. The more rich and diverse that creation, the more God's goodness is reflected, 'for in knowing the excellence of the world we may speak analogically about the One in whose being it participates.'[24] Johnson goes on to establish that we cannot denigrate the merely natural, when, in faith, we understand that it exists due to its participation in the fullness of God's life. In this way our faith tradition can strengthen our resolve to address the critical issues facing Earth and

---

20 Ibid., p. 147.
21 Clarke, *Explorations in Metaphysics*, p. 265.
22 Johnson, *Ask the Beasts,* p. 147.
23 Ibid., p. 148.
24 Ibid., p. 149.

cosmos that are explored under Principle 1.

The natural world, the entangled bank, is sacred. It is imbued with a spiritual presence that holds it in existence. It is the dwelling place of God. In Johnson's words, 'Earth is a physical place of extravagant dynamism that bodies forth the gracious presence of God. In its own way it is a sacrament and a revelation.' Its materiality embodies 'the active presence of the holy Giver of life,' an understanding that is acted out in the sacraments of the Church, itself 'constituted as a sacrament of Christ's presence for the world.'[25] Earth Link can be sure that matter does reveal, in the biblical sense of teaching us about God. It can be seen as a book that reveals in much the same way as does Scripture, an image favoured by Augustine[26], who exhorts people to observe Heaven and Earth and see the works of God. In similar vein, the poet Hopkins observed and exclaimed that 'the world is charged with the glory of God.'[27] This motif is present in *Laudato Si*, where we are reminded that 'God has written a precious book … a constant source of wonder and awe.'[28]

Awareness of nature as revelatory is not new in the Christian tradition, and it continues to be richly developed in the thought of Pope Francis in *Laudato Si.* This is superb prose and rich food for ecospiritual reflection:

> The entire material universe speaks of God's love, his boundless affection for us. Soil, water, mountains, everything, are, as it were, a caress of God. The history of our friendship with God is always linked to particular places which take on an intensely personal meaning; we all remember places, and revisiting those memories does us much good. For anyone who has grown up in the hills or used to sit by the spring to drink, or played outdoors in the neighbourhood square, going back to those places is a chance to recover something of their true selves.[29]

---

25 Ibid., p. 151.
26 Augustine, 'Sermon 68:6', pp. 225–6.
27 Gerard Manley Hopkins, *God's Grandeur,* https://www.poets.org/poetsorg/poem/gods-grandeur
28 Pope Francis, *Laudato Si,* p. 85.
29 Ibid., p. 84.

And there is more. Edwards summarises the references in the encyclical to the relation between God and all that is not God:

> Other creatures 'speak' to us of God's love, are a 'caress' of God, a 'precious book' whose letters are the multitude of created things, a 'manifestation' of God, a 'continuing revelation' of the divine, a 'teaching' that God wishes to give us, a 'message' from God, and a 'divine manifestation.'[30]

Alongside revelation from Scripture and tradition, which Johnson richly presents, here is revelation from nature, before which we need to be open, attentive and aware. Ecospirituality has a distinctive faith content.

The emerging universe is classified by science as natural. Faith encourages us to see it as the dwelling place of God, who created it and sustains it through the presence of the Spirit. It is sacred, and worthy of our respect, reverence and care. The emerging universe is now known to be unfolding according to its own dynamics, and that raises new questions about the way in which God acts in the universe. Earth Link needs to address this issue if it takes evolutionary emergence seriously.

Our deep bonding with the community of creation engages us in listening to diverse voices and their searches for God at the edges of established power, a pursuit dear to Elizabeth Johnson. In listening to Darwin's tangled bank and its evolutionary story she poses a new question that takes her beyond the theology of God's presence, which up to this point could equally apply within a static world view. She asks, 'If indeed the current design is the result of a long history that can be explained by natural laws known to us, how are we to understand not just the presence but the activity of God?'[31] The universe seems to be emerging through its own innate powers, yet Job's reflection on the beasts reminds us that what is happening is due to the hand of God. How do these perspectives go together? This is a question that Earth Link also needs to address, as much popular spirituality presumes a static view of the natural world.

---

30 Edwards, "'Sublime Communion'", p. 384.
31 Johnson, *Ask the Beasts,* p. 154.

The action of the God of love

God as monarch has been the dominant model for the God who rules directly over his subordinates and nature such that they predictably fulfil his design. Natural selection poses a significant disruption to that model because it demonstrates empirically that variations have been occurring over eons of time. As Johnson points out, '[T]he absence of direct design, the presence of genuine chance, the enormity of suffering and extinction and the ambling character of life's emergence over billions of years are hard to reconcile with a simple monarchical idea of the Creator at work.'[32] Hence she poses a new dilemma, which is about 'how to understand the presence of the Spirit of God acting continuously to create in the light of evolutionary discoveries about the entangled bank.'[33]

The theological interpretation that she offers is that 'God's creative activity brings into being a universe endowed with the innate capacity to evolve by the operating of its own natural powers, making it a free partner in its own creation.'[34] This is a model based on love rather than power over. This is no deistic God who creates and then leaves the world to its own devices. This is no monarchical God who manages every detail. It is rather that the Giver of life 'freely and generously invests nature with the power to organise itself and emerge into ever-new, more complex forms, and to do so according to its own ways of operating'.[35] Nature is always being called forward to newness while being empowered from within. This is evolutionary science taken into a theological framework, which is what we are doing in the Principles.

Johnson goes on to understand God's action in the natural world as an extension of our understanding of God's saving action throughout history. She draws on Rahner's position that 'if we see the created world emerging thanks to the self-giving love of God,'[36] the most helpful framework is that of grace, or God's graciousness. God's action

---

32 Ibid., p. 155.
33 Ibid.
34 Ibid.
35 Ibid., p. 156.
36 Ibid.

in Christ through the Spirit is currently the model of God's action in creation. Johnson draws on various sources in the tradition that argue that God's action in human beings enhances rather than detracts from their existence. Irenaeus' famous saying is that 'the glory of God is the human being fully alive.'[37] God's glory and human vitality are directly correlated.

Johnson draws on insights about grace as experienced by humans and extends them to all species. She infers that

> [t]he belief that God is faithful and acts consistently provides a warrant for thinking that as with humans, so too, with the natural world from which we have evolved. The gracious God, Spirit proceeding as love in person, is present to bless and enhance natural powers rather than to compete with them. With such a love there can be no anxiety about control.[38]

This love is embodied in the life, death and resurrection of Jesus understood as 'compassionate self-giving love for the liberation of others'[39], a love that flows to all. This is about understanding God's omnipotence as empowering rather than overpowering. It is love that enables the flourishing of the other and its autonomous development. God who dwells in the midst of creation empowers nature to work towards its own fullness. This allows for the emergence of more complex life forms out of less complex ones. Summarising this paradigm of love, Johnson says, 'In a theological perspective, this whole process is empowered by the Creator who as love freely gifts the natural world with creative agency. Its relation to the living God is marked simultaneously by ontological dependence and operational autonomy.'[40]

The action of the God of love enables natural processes, and the natural world enters into the enabling process with its own integrity. This is the God we encounter in Earth and cosmos.

---

37 Irenaeus, *Adversus Haeresis,* 4.20.7; see also 3.20.2 and 5.3.
38 Johnson, *Ask the Beasts,* p. 158.
39 Ibid.
40 Ibid., p. 160.

Divine and created agency in an evolving world

Earth Link acknowledges that God not only dwells within Earth and cosmos, but also exercises agency. God's action in evolutionary emergence is an area of considerable theological discussion, and a brief survey of such is helpful within the scope of this principle. It is also an issue of importance to all of us who invoke God's aid in our personal and global crises. Johnson indicates that 'a large cadre of theologians endorse the idea of nature's independent working as shown by evolution. They diverge mightily, however, over how to think about the relation between divine and created agency in an evolutionary world.'[41] Earth Link needs to engage with these discussions. Johnson gives a snapshot of a range of positions: single action theory (e.g. Gordon Kaufman et al.), top-down causality (Arthur Peacocke), causal joint theory (e.g. Polkinghorne), organic model of world as body of God (e.g. Sallie McFague), kenotic self-emptying love (e.g. John Haught), and process thought (e.g. John Cobb).[42] Despite their considerable differences, they are all seeking to understand how 'the creating God as ground, sustaining power, and goal of the evolving world acts by empowering the process from within' while respecting 'the freedom of the natural world to evolve consistently with it internal laws as discovered by contemporary science.'[43]

Johnson herself espouses the scholastic position of primary and secondary causality, whose basic principle is that 'the creative activity of God is accomplished in and through the free working of secondary causes.'[44] She argues that this position is still appropriate within a dynamic evolutionary world view because it protects the distinction between Creator and created while allowing for a sense of co-creation. Johnson first establishes the distinction between an ultimate cause and a proximate cause. One is the 'Cause of all causes,' and the other participates in 'the power to act, as things that are burning participate

---

41 Ibid.
42 Ibid., pp. 161–2.
43 Ibid., p. 163.
44 Ibid.

in the power of fire.'[45] She draws on Aquinas' argument for the agency of creatures according to their own nature. In summing up, she says, 'It is characteristic of the creative power of God to raise up creatures who participate in divine being to such a degree that they are also creative and sustaining in their own right.'[46] Just as creatures participate in God's being and goodness, they participate in God's agency. God's agency in creation gives creatures their own autonomous agency. God's exercise of ultimate causality does not fill the gaps of any secondary or creaturely action:

> God's act is not a discrete ingredient that can be isolated and identified as a finite constituent of the world. In this sense the world necessarily hides divine action from us. The living God acts by divine power in and through the acts of finite agents which have genuine causal efficacy in their own right.

Johnson then discusses divine governance of the world, in which primary and secondary causality are related to final causation, understood as 'a creature's innate tendency toward a goal.'[47] This self-direction, integral as it is to creatures, is a natural inclination toward good and ultimately toward participation in divine goodness. In scholastic terms, God's immanence is present as final cause, or in biblical terms, it is 'reaching from one end of the world to the other' (Wis 8:1).

According to Johnson, this understanding of God as immanent Ground of all Being is consonant with the evolutionary understanding of the autonomy of all that is. God as ultimate cause endows beings with their participation in divine being, agency, and goodness, which they then act out. God exercises power by giving causative power to others. God is not just another secondary power. Rather, 'the Spirit of God continuously interacts with the world to implement the divine purpose by granting creatures and created systems their full measure of efficacy.'[48]

Johnson asserts that this dynamic is effective only within the context

---

45 Ibid.
46 Ibid., p. 164.
47 Ibid., p. 165.
48 Ibid., p. 166.

of the 'overarching notion of the Creator God as the absolute Living One, pure wellspring of being, and the concomitant notion of creaturely participation.'[49] For her that overarching conviction allows the world to evolve in its own way while manifesting God's wisdom.

Another approach with merit in the consideration of God's action is the 'whole-part' and 'top-down' model of Peacocke. In the case of emergent reality, the whole and the parts have different causal powers that are interrelated but independent. The whole transcends the parts, yet can have a downward influence. Peacocke is careful to avoid any suggestion of direct divine intervention in specific aspects of the natural evolutionary process:

> By analogy with the operation of the whole-part influence in natural systems, I have in the past suggested that, because the 'ontological gap/s' between the world and God is/are located simply *everywhere* in space and time, God could affect holistically the state of the world (the whole in this context) at all levels. Understood in this way, the proposal implies that patterns of events at the physical, biological, human and even social levels could be influenced by divine intention without abrogating natural regularities at any of these levels.[50]

This analogy is limited to the relationship between God's intention and the whole system, which in turn influences the parts or various levels. Peacocke also admits the limits of his approach:

> I hope that the model described above so far has a degree of plausibility in that it depends only on an analogy with complex natural systems in general and on the way whole-part influence operates in them. It is, however too impersonal to do justice to the *personal* character of many (but not all) of the profoundest experiences of God.[51]

The debate goes on. However, as a basis for ecospirituality, it is timely to recognise, with Peacocke, that 'because of the "ontological gap/s" between God and the world which must always exist in any theistic model, this is only an attempt at making intelligible that which

---

49 Ibid., p. 168.
50 Peacocke, 'Emergence, Mind', p. 275.
51 Ibid., p. 277.

we can postulate as being the initial effect of God experienced from, as it were, our side of the ontological boundary.'

So what about the emergence of the new, of those chance occurrences which seem to be a mark of the evolutionary process? Biologically there is an interplay of law and chance. Clarifying the meaning of these terms, Johnson says, 'Law refers to an orderly suite of natural forces that govern how the universe works. These principles, read off from the regularities observed in the world, hold true in all ordinary circumstances.'[52]

Law accounts for predictability in what could otherwise be chaos. Law is not pre-ordained but based on description and analysis of what is happening. Chance, on the other hand, refers to 'the crossing of two independent causal chains that intersect for no known reasons that can be figured out in advance.'[53] This is the realm of the unexpected, of open-endedness. When law and chance work together there is scope for newness with some constraint, such as when natural selection deems that variations more suited to the environment thrive while other variations do not endure.

When it comes to theological reflection on the interplay of law and chance, Johnson indicates that 'theology has traditionally allied God with lawful regularity,'[54] with having a plan and a purpose. There are definitely laws in nature that can be understood to reflect 'the faithfulness of the living God, reliable and solid as a rock.'[55] However the evolutionary emergence of the new, apparently by chance, can be seen as reflecting 'the infinite creativity of the living God, endless source of new possibilities.'[56] This new emphasis on creativity points to the way in which 'the natural world participates in its own creation.'[57] The Creator Spirit is present in these processes. As Johnson puts it, 'Divine Love empowers the structure of creation which operates with its own integrity, all the while supporting unfolding events as they

---

52 Johnson, *Ask the Beasts,* p. 169.
53 Ibid., p. 170.
54 Ibid., p. 173.
55 Ibid.
56 Ibid., p. 174.
57 Ibid.

weave into regular patterns toward the realization of an ever more complex whole.'[58]

Evolutionary emergence has opened up new understandings of matter. Evolutionary theory traces the progression from 'inanimate to animate to intentional states, emerging into greater complexity from within.'[59] In that process a new biological form can emerge that 'gathers up what has preceded it, shaping this material into a more complex unity. What emerges has distinctly different properties and functions from what went before, though [it is] still composed of the same fundamental matter.'[60] It transcends yet includes what went before.[61] Such understandings challenge the view of matter as inert and static, an attitude typical of philosophical dualism, in favour of its emergent properties. Remember that this is a foundational position of Earth Link.

Turning to theology, Johnson refers to Rahner, who speaks of matter's 'capacity to transcend itself.'[62] Rahner considers first the dynamic unity of matter and spirit in the human person. Body, mind and spirit act as one. By extension, Rahner suggests that 'matter develops out of its own inner being in the direction of spirit.'[63] This capacity for self-transcendence is innate, not something added on. He understands this as a process of becoming, whereby nature 'reaches an inner increase of being proper to itself … and does this not by adding something on but from within.'[64] Rahner takes this insight into an understanding of divine presence as something so intrinsic to the creature that 'the finite being is empowered by it to achieve a really *active* self-transcendence and does not merely receive this new reality passively effected by God.'[65] Matter can emerge into life and into spirit, into a relationship with Divine presence active within.

---

58 Ibid., p. 175.
59 Ibid., p. 174.
60 Ibid.
61 Phipps, *Evolutionaries*, p. 191. Phipps acknowledges Hegel's recognition of the pattern of 'transcending and including' as a universal principle of evolution. I would add that it is also a pattern of emergence.
62 Rahner, 'Christology within an Evolutionary View of the World', p. 164.
63 Ibid., p. 165.
64 Ibid., p. 175.
65 Ibid, p. 162.

Johnson calls to mind those theologians who are exploring artistic metaphors to capture something of the relation of the Giver of Life to the cosmos unfolding autonomously from within. The Creator Spirit is likened to the composer of a fugue who builds up a composition from a theme that is embellished in counterpoint. The Spirit can be likened to a jazz musician who improvises spontaneously on a theme, or an improviser in a dramatic performance. Similarly, the Spirit is likened to a choreographer creating in collaboration with the dance troupe, or the designer of a game of cards. Gone is the controlling ruler. Rather the quest is for 'an understanding of faith that renders fair account of the intense creative activity of both Creator and creation.'[66] For Johnson, a theology of the Spirit contributes much to this quest. For her, 'infinite mystery of self-giving love, the Creator Spirit, calls the world into being, gifts it with dynamism, and accompanies it through the by-ways of evolution, all the while attracting it toward a multitude of "endless forms most beautiful".'[67] The action of the Spirit is based on loving respect for creaturely freedom and autonomy.

Johnson returns to the biblical images of the Spirit − blowing wind, flowing water, burning fire, brooding bird, holy Wisdom − who moves with the dynamic and creative community of life. Job's faith vision when he looks at the beasts and the diversity of Darwin's tangled bank can work together when we acknowledge the intrinsic value of the natural world in its own right recognise it as the dwelling place of God. Similarly, ecospirituality can foster 'respect, reverence and care for the Earth community.'[68]

Recognition of the intrinsic value of all reality is a new insight for official Catholic thought and receives vigorous attention in *Laudato Si*. In Catholic thought up to now, we have been exhorted to care for creation because of its importance to humans, a social justice perspective with merit, but one that is not inclusive of the whole of reality. Pope Francis exhorts people to respect and protect ecosystems. They are

---

66 Johnson, *Ask the Beasts,* p. 178.
67 Ibid.
68 See enhanced vision of Earth Link, https://www.earth-link.org.au/

created in love.[69] Humans do use them, but that use is tempered by a new awareness:

> We take these systems into account, not only to determine how best to use them, but also because they have an intrinsic value independent of their usefulness. Each organism, as a creature of God, is good and admirable in itself; the same is true of the harmonious ensemble of organisms existing in a defined space and functioning as a system.[70]

Ecosystems are valuable because 'the Spirit of God dwells in them'[71] and because they are the locus of our encounter with God. In them, Heaven and Earth embrace. In acknowledging Earth and cosmos as the dwelling place of God, Earth Link recognises that it is indeed God whom we encounter in sacred Earth if such is our faith perspective. God brings all into existence by creating in love. God's Spirit dwells within all, continuously creating, without denying the autonomy of what is created. God's action as we experience it enables us to trust that the community of life is also moving forward toward fullness in God, a conviction that is consolidated as we acknowledge that the God of love embraces the cosmos in healing solidarity in Jesus Christ.

---

69 Pope Francis, *Laudato Si,* p. 76.
70 Ibid., p. 140.
71 Pope Francis, *Laudato Si*, p. 88.

CHAPTER 7

Principle 4: Jesus In Solidarity

Acknowledge that in Jesus Christ the God of love embraces the cosmos in healing solidarity.

This principle is pivotal to any ecospirituality that acknowledges the centrality of Jesus Christ in its faith tradition, and that attempts to make meaning of the evolutionary realities of pain, suffering and death. Jesus embraces fully our finite cosmos, revealing the loving heart of God. In his life, death and resurrection, Heaven and Earth embrace in a very special way, conveying a message of hope and the promise of fullness of life with God in the future. Jesus Christ endures pain, suffering and death in solidarity not only with humans, but with the whole created reality. The evolutionary story testifies to the presence of pain, suffering and death as a necessary condition for the survival of the species and transition to new forms. In other words, the evolutionary story is marked by breakdown as well as creativity. Darwin's work on common descent with modification reminds us that evolution comes at a cost. In the evolutionary story, pain, suffering and death are not the result of any particular human actions, although undoubtedly human beings experience them with heightened sensitivity. The relations between humans and the environment are the subject of Principle 5.

Jesus Christ embraces our finite habitat. Johnson's Trinitarian theology explores the deep incarnation and deep resurrection of Jesus Christ. In particular, she elaborates on the consequences of his fleshiness within the cosmos and Planet Earth, an identification that offers a meaningful framework for our own and nature's experience of pain, suffering and death. This is an important perspective for an ecospiritual way of life.

Jesus as creative, revealing and saving presence of God in the world

The post-Resurrection community had the unique experience of knowing Jesus in the flesh, witnessing his horrendous death, and experiencing his presence in a new way after his resurrection. They identified him as Emmanuel, God-with-us. In identifying Jesus with the Hebrew personification of Wisdom, they linked him with the 'creative, revealing, and saving presence of God'[1] acting in the world. The Johannine community recognised him as the Word, as the 'coming of God's personal self-expressing Word, full of loving kindness and faithfulness, into the world.'[2] The Word was creative and life-giving, in continuity with the Genesis narrative, and that Word assumed flesh in the person of Jesus. This self-utterance of God became flesh and lived among them. This was not a mere appearance, but as Johnson puts it, 'Taking the ancient theme of God's dwelling among the people of Israel a step further, it affirms that in a new and saving event the Word *became* flesh, entered into the sphere of the material to shed light on all from within.'[3]

Creator and created are radically connected in the person of Jesus. Jesus in turn is radically connected to the stuff of the universe. This is Gregersen's 'deep incarnation'.[4] Jesus bears 'the signature of the supernovas and the geology and life story of the Earth.'[5] Jesus is bonded, not only to humans whose form he took, but to the universe that gave rise to his matter. In the incarnation, Jesus assumes the universe, not just human flesh.

This understanding is a distinct contribution of Christianity, given its core belief in the divinity and humanity of Jesus. The notion that 'the one transcendent God who creates and empowers the world freely chooses to join this world in the flesh, so that it becomes a part of God's own divine story forever'[6] impacts not only on our understanding of

1 Johnson, *Ask the Beasts,* p. 193.
2 Ibid., p. 194.
3 Ibid., p. 195.
4 Ibid., p. 196.
5 Gregersen, 'Cross of Christ in an Evolutionary World', pp. 192–207.
6 Johnson, *Ask the Beasts,* p. 197.

God, but also on the universe. The incarnation has implications for all created reality, which is united with God in a new way. Pope John Paul II puts this very clearly:

> The Incarnation of God the Son signifies the taking up into unity with God not only of human nature, but in this human nature, in a sense of everything that is 'flesh': the whole of humanity, the entire visible and material world. The Incarnation, then, also has a cosmic significance, a cosmic dimension. The 'first-born of all creation', becoming incarnate in the individual humanity of Christ, unites himself in some way with the entire reality of humanity – which is also 'flesh' – and in this reality with all 'flesh', with the whole of creation.[7]

Jesus became incarnate in the whole of creation, 'harsh, perilous, mighty, universal, impenetrable, and mortal though this material stuff be.'[8] Johnson reminds us that Teilhard de Chardin appreciated all of matter as a divine milieu 'charged with creative power, as the ocean stirred by the Spirit, as the clay moulded and infused by the Incarnate Word.'[9] This deep incarnation of God in matter is a vital union, with many consequences for engaging spiritually in an emerging universe.

Revealing the heart of God

In Jesus Christ we gain remarkable insight into the way in which God relates to the world. We learn that God's creative dream for the world is indeed good news for all, especially the lost, the lowly, even the lilies of the field. He worked for the wellbeing of all reality in a way that integrated body, mind and spirit. In the words of McFague, 'liberating, healing and inclusive love is the meaning of it all,'[10] and that love is inclusive of all parts of any ecosystem.

Jesus paid a very high price for this commitment. He suffered an excruciating death at the hands of his captors. As distinct from the suffering and death that is integral to the evolutionary process, the death of Jesus was 'a contingent event resulting from an expedient decision

---

7 Pope John Paul II, *Lord and Giver,* p. 50.

8 Johnson, *Ask the Beasts,* p. 198.

9 Teilhard de Chardin, *Hymn of the Universe,* p. 70.

10 McFague, *Body of God,* p. 161.

by political authorities. Far from being the result of a natural process, the crucifixion was historical, unpredictable, unjust, the result of human sin.'[11] Here, the relation between the Creator God and creation is expressed in a new way. God's immersion in matter involved dying, a 'seemingly non-godly characteristic.'[12] Paul in Philippians saw Jesus Christ as one who 'emptied himself, taking the form of a slave, being born in human likeness ... and became obedient to the point of death, even death on a cross'(Phil 2:7). This is a far cry from the model of an omnipotent and distant monarch. Instead, it reveals a God who is 'self-emptying, self-limiting, vulnerable, self-giving, in a word, creative Love in action.'[13]

In the biblical tradition, there is a strong sense of God's compassionate presence with the suffering creation. As the Israelites wandered in the desert after their escape from oppression in Egypt, God assured Moses that he 'knew' of their suffering (Exodus 3:7). Johnson points out that 'what is new in view of the cross is divine participation in pain and death from *within* the world of the flesh. Now the incarnate God knows through personal experience, so to speak.'[14] She sums this up very concisely: 'God suffers.'[15]

Classical theology has long sought to hold together an understanding of the human and divine natures in the person of Jesus, which they have expressed as a hypostatic union. Johnson draws on the more contemporary theology of Walter Kasper, who points to the unexpected, shocking death of Jesus on the cross as 'the unsurpassable self-definition of God.'[16] This suffering, freely chosen, is an expression of love.

It is an important insight for Earth Link that humans in the intensity of their suffering find here a God with whom they can identify, a God who cries out on the cross, 'My God, my God, why have you forsaken me?' (Matt 27:46, 50). Drawing on Jürgen Moltmann, Johnson points out that '[i]t is as if by inhabiting the inside of the isolating shell of

---

11 Johnson, *Ask the Beasts,* p. 202.
12 Ibid.
13 Ibid.
14 Ibid., p. 203.
15 Ibid.
16 Kasper, *God of Jesus Christ*, p. 194.

death, Christ crucified brings divine life into closest contact with disaster, setting up a gleam of light for all others who suffer in that same annihilating darkness.'[17]

This is both a personal and a political message. Liberation theologians see in the death of Jesus an expression of solidarity with others who die by the edict of political leaders. It is also an ecological message. Johnson again draws on Gregersen's understanding of deep incarnation:

> The logic of deep incarnation gives a strong warrant for extending divine solidarity from the cross into the groan of suffering and the silence of death of all creation. All creatures come to an end; those with nervous systems know pain and suffering. Jesus' anguished end places him among this company.[18]

In Jesus, God embraces the evolving world and the inevitability of pain and death, or in the words of *Laudato Si*, '[O]ne Person of the Trinity entered into the created cosmos, throwing in his lot with it, even to the cross.'[19]

In solidarity with groaning creation

Jesus Christ came with a message of liberating, healing love. He paid the ultimate price with his life. This does not make a virtue of suffering, and as with Johnson, Earth Link does not justify suffering. Johnson has given serious consideration to this issue as it has been developing in the wake of the Holocaust disaster. In *She Who Is*, Johnson considered the mission, suffering and death of Jesus-Sophia, indicating that evil does not have the final word. She rejects any justification of suffering in favour of a theology of solidarity, insisting that '[w]hat comes clear in the event, however, is not Jesus' necessary passive victimisation divinely decreed as a penalty for sin, but rather a dialectic of disaster and powerful human love through which the gracious God of Jesus enters into solidarity with all those who suffer and are lost.'[20]

In *Ask the Beasts,* Johnson needs to go further to make meaning of

---

17 Johnson, *Ask the Beasts,* p. 204.
18 Ibid, p. 205.
19 Pope Francis, *Laudato Si*, p. 99.
20 Johnson, *She Who Is,* p. 156.

Paul's observation in his letter to the Romans: 'We know that the whole creation has been groaning in labour pains until now' (Rom 8:22). Paul includes the whole of suffering creation in the promise of the 'freedom of the glory of the children of God' (Rom 8:21).

Johnson notes that biologically, the emergence of sentience also meant the emergence of the capacity for experiencing pain as well as forms of heightened awareness that can serve to alert species to danger. The greater the sensitivity, the greater the capacity for suffering and pleasure. Fossil records bear witness to the extinctions not only of individuals, but sometimes of whole species. Natural selection means that some less well-adapted species come to an end, creating more opportunity for the better adapted. Johnson confines her observations to 'extinctions that happen spontaneously in nature apart from the actions of the human species.'[21] Earth Link will take up human actions in Principle 5. The extinctions Johnson is considering are morally neutral. Johnson notes the inevitability of such pain and death: 'The laws known to us which have brought about the entangled bank, so pleasing in its beauty, have also rendered it a place of pain and death.'[22]

Johnson goes on to frame theologically the conversation that is needed about this reality. As she did with the question of human suffering, Johnson rejects theodicy, which seeks to 'construct a rational defense of God's goodness and power in a world where evil occurs.'[23] Such justification leads to passive acceptance of enormous suffering, and of the political and economic situations that often give rise to it.[24] Rather, she seeks 'a theological inquiry that takes the evolutionary function of affliction at face value and seeks to reflect on its workings in view of the God of Love made known in revelation.'[25] Johnson stays with what she has learned from her study of Darwin and seeks to address evolutionary suffering independent of human ethical responsibilities. In the process

---

21 Johnson, *Ask the Beasts*, p. 185.

22 Ibid., p. 186.

23 Ibid., p. 187.

24 Delio suggests that the real theodicy question is 'not why God allows bad things to happen to good people, but why we abandon God in the face of suffering.' See Delio, *Unbearable Wholeness*, p. 83.

25 Johnson, *Ask the Beasts,* p. 187.

she differentiates herself from the likes of Celia Deane-Drummond and John Haught, who are concerned to engender resistance to, rather than passive acceptance of, the inevitability of suffering and death in creation. Johnson acknowledges that such resistance is very important but underlines that she wishes to begin her explorations by listening to the beasts, by acknowledging 'the finite character of the natural world and ... its role in the evolutionary process.'[26] She affirms this as an expression of the autonomy of natural processes.[27] From that starting point she indicates that 'the most fundamental move theology can make ... is to affirm the compassionate presence of God in the midst of the shocking enormity of pain and death.'[28] God in the Hebrew scriptures is a God of pathos, at times delighted and at times full of lament for the devastation of land and people. The Christian tradition is built on the life, suffering, death and resurrection of Jesus Christ, one who entered into the beauty and the pain of an Earthly existence, to the point of dying a tortuous death. However, in the experience of encountering the resurrected Christ, the tradition came to the belief that the death of Jesus was not the end, but, indeed, part of a new creation that forms the basis of hope. There is a vital connection between the groaning of creation and the God of Love, God who in Jesus Christ is in solidarity in the midst of suffering.

In the encyclical, *Laudato Si*, Pope Francis also shows keen awareness of the groaning of creation, but whereas Johnson is intent on 'listening to the beasts', his focus is on the Earth-human interaction, and he quickly looks at human causes even as he reminds us of our kinship with groaning Earth:

> This sister now cries out to us because of the harm we have inflicted on her by our irresponsible use and abuse of the goods with which God has endowed her. We have come to see ourselves as her lords

---

26 Ibid., p. 191.

27 The autonomy of created reality is also supported in *Laudato Si*: 'One Person of the Trinity entered into the created cosmos, throwing in his lot with it, even to the cross. From the beginning of the world, but particularly through the incarnation, the mystery of Christ is at work in a hidden manner in the natural world as a whole without thereby impinging on its autonomy' (p. 99).

28 Johnson, *Ask the Beasts*, p. 191.

and masters, entitled to plunder her at will. The violence present in our hearts, wounded by sin, is also reflected in the symptoms of sickness evident in the soil, in the water, in the air and in all forms of life. This is why the earth herself, burdened and laid waste, is among the most abandoned and maltreated of our poor; she 'groans in travail' (Rom 8:22). We have forgotten that we ourselves are dust of the earth (cf. Gen 2:7); our very bodies are made up of her elements, we breathe her air and we receive life and refreshment from her waters. *Nothing in this world is indifferent to us.*[29]

There was and is more to the experience of Jesus in his identification with material reality. There must be, or we would not be talking about the Christ event some centuries later. Johnson does not espouse any notion that Jesus' resurrection was a return to his former biological state, yet it conveys a message about physicality. The followers of Jesus experienced him as risen, not only in spirit but also in his body. They had a strong sense of encounter with him after his death, and recognised him as the 'firstborn of all creation' (Col 1:15). Death is not the whole of the story, either for Jesus or for our valuable Earth. The Pauline community understood death as the beginning of something new, as the promise of fullness of life.

Johnson extends Gregersen's deep incarnation into deep resurrection, which, as she says, 'extends the risen Christ's affiliation to the whole natural world.'[30] The risen Christ takes corporeality in all its manifestations into the heart of God. Christ entered freely into the emerging universe, experienced its limits, gave of himself unto death, and moved beyond that into fullness of life in God. Johnson draws attention to the moment in the Easter liturgy, at which we celebrate that all creation shares in the transition from darkness to light:

> Rejoice, O earth, in shining splendour,
> Radiant in the brightness of your King!
> Christ has conquered! Glory fills you!
> Darkness vanishes forever!

Death is the preface to life and to a new mode of relationship for

---

29 Pope Francis, *Laudato Si*, p. 2.
30 Johnson, *Ask the Beasts*, p. 208.

Christ, for us and for all creation. As Pope Francis says in *Laudato Si,*

> [t]he creatures of this world no longer appear to us under merely natural guise because the risen One is mysteriously holding them to himself and directing them towards fullness as their end. The very flowers of the field and the birds which his human eyes contemplated and admired are now imbued with his radiant presence.[31]

Or, in the words of Delio, '[F]inite life is released from its limits to become part of something more than itself, a new whole of cosmotheandric life, new relatedness with God and cosmos that we name the risen Christ.'[32] We have a source of hope for the whole community of life in Christ's solidarity with the emerging universe. These are vital perspectives for Earth Link as part of its publically Catholic Christian profile.

In the beginning

In terms of exploring the theological underpinnings of ecospirituality within a Judaeo-Christian context, it might be expected that one would begin with creation narratives. Instead, in this principle, Earth Link follows Johnson and begins with the presence of the Spirit in cosmos and Earth in the act of continuous creation, then considers Jesus Christ as further revealing the heart of God in the process of assuming flesh and dwelling among us. These are observable phenomena. How do we think and theologise about issues that are not open to observation? These are beyond the field of science, which postulates about how things began and where they are going, and is continually making advances. This is, rather, the field of theology, which engages with the how and the why and where to, based on 'the living tradition's knowledge of God's graciousness given through Jesus in the power of the Spirit.'[33] Faith assertions emerging from the lived experience of grace in the Christian community are extended to beginnings and endings. They are not based on empirical data, but rather 'predicated backward and forward beyond

---

31 Pope Francis, *Laudato Si*, p. 100.
32 Delio, *Unbearable Wholeness*, p. 201.
33 Johnson, *Ask the Beasts,* p. 212.

time, to where no experience can go.'[34] The strength of such assertions comes from the sense that 'as the living God is now, so God was and will be.'[35] The Jewish people believed that the God who liberated them in the Exodus, and who would redeem them when in exile, was the one who created all that is, and who would be faithful into a blessed future. Johnson adopts this 'hermeneutic of the present experience of grace'[36] to link the God of the beginning with the God of the end, and make the case for the action of Holy Mystery in creating life and working towards its ultimate fulfilment beyond death. This is a very helpful perspective for Earth Link in its approach to creation theology and eschatology.

In *Laudato Si,* Pope Francis points to the biblical understanding that 'the God who liberates and saves is the same God who created the universe, and these two divine ways of acting are intimately and inseparably connected.'[37] His context is the chapter called 'The Gospel of Creation.' Johnson's context is her Trinitarian theology, which is in dialogue with the evolution of species in the light of the theories of Darwin. She begins with continuing creation (*creatio continuo*), as she said this best suited her dialogue with Darwin and the beasts. She goes on to look at creation 'beyond time'[38], namely, in the beginning (*creatio originalis),* and in the new creation (*creatio nova).*

Johnson turns to Job's message from the beasts that they are created. They 'do not explain or ground themselves, but are brought into being by God's creative word.'[39] Creation is a gift from the Giver of life. Matter itself is not divine.[40] Johnson reads the creation narratives of the first two chapters of Genesis with a focus on the beasts, birds, plants and fish, and identifies the religious message that 'God is the Maker of heaven and earth and all that is in them.'[41] Later, when Christianity encountered the Graeco-Roman world, it distinguished itself from

---

34 Ibid.
35 Ibid.
36 Ibid., p. 213.
37 Pope Francis, *Laudato Si*, p. 73.
38 Johnson, *Ask the Beasts,* p. 212.
39 Ibid., p. 215.
40 This point is also clearly made in *Laudato Si*, p. 78.
41 Johnson, *Ask the Beasts,* p. 215.

pantheism and Gnosticism with the belief summed up in the phrase 'creatio ex nihilo', creation from absolutely nothing. Johnson considers three possible meanings of creation from nothing. The first is that there is only one source of all that is, namely, Infinite Mystery. The second is that creation from nothing signals the goodness of all things, including material creatures. It signals, the third argument asserts, that the existence of the world is a free gift – not a necessity, but an act of God's own gracious, loving will.[42] Reality, as created, is God's good gift, given in love.

Contemporary evolutionary science challenges any notion that creation happened literally in a week as depicted in the mythic narratives of Genesis. Knowledge about the emergence of the universe, currently dated at around 13.7 billion years ago, does not challenge the basic premise of Genesis that 'whatever the manner in which the present world came into being, it would still be ontologically grounded in God's creative act.'[43] This grounding is also a feature of God's continuous creation of both universe and species. The emergence of life around four billion years ago and the subsequent flourishing of species find their ultimate ground in 'the God of Love as creating Source.'[44] In the Judaeo-Christian tradition, the God who created does not use creation to wreak havoc, despite the reality of drought, floods and fires. God entered into a covenant relationship with 'every living creature of all flesh that is on the earth'(Gen 9:16), which is an assurance of God's presence and God's remembering from the beginning and into the future.

At the end

As Johnson considers *creatio nova* (the new creation) at the end of time, she faces squarely some of the scientific hypotheses about the inevitability of the universe, the sun, and species coming to an end, and she distinguishes scientific from religious perspectives. Scientifically there is uncertainty about whether the universe is on track to an implosion or an explosion. It is agreed that the sun is half-way through its life and that its potential

---

42 Ibid., p. 216.
43 Ibid., p. 218.
44 Ibid., p. 208.

development into a red giant, maybe in about five billion years, will bring about the end of our planet and life as we know it.

Religious thought has had a preoccupation with the end times since the first century, and it has given rise to apocalyptic language and scenarios. This preoccupation does contain a message of hope, despite signs to the contrary. As noted above, Johnson highlights that such future scenarios are extrapolated from current knowledge and experience. She sums it up this way: 'What such speech does do is affirm the core conviction that all of reality exists within the embrace of God's gracious love, and that it is going toward a fulfilment yet to come.'[45] Just as the creation story places reality in God's hands, the end times are about a return to the fullness of the love that originated, sustained, and will draw reality into wholeness beyond its finite existence.

Johnson turns next to cosmic redemption, considering the inclusion of all of creation in the redemptive process. The motifs of 'atonement, satisfaction and sacrifice'[46] have dominated theological understandings for some time. Johnson traces the origins of this preoccupation to Anselm in the eleventh century, although she recognises that in his hands it was about recognising God's mercy in making satisfaction for the sinfulness of finite humans. This preoccupation descended into an exaltation of 'Jesus' necessary passive victimisation divinely decreed as a penalty for sin.'[47] Our primary identity became that of sinner, an identity that still dominates the language of liturgical practice. It is an identity that many who are interested in ecospirituality seek to cast off, often without knowing what to replace it with. Johnson points to developments during the last century that instead enshrine the centrality of 'the mystery of grace poured out in the crucified and risen Jesus,'[48] using concepts such as 'liberation, reconciliation, healing, justification, victory over the powers, living in peace, fullness of life, being freed from slavery, adoption and new birth as God's children'[49] as ways of interpreting the centrality of the experience of redemption.

---

45 Ibid., p. 221.
46 Ibid., p. 223.
47 Johnson, *She Who Is*, p. 158.
48 Johnson, *Ask the Beasts*, p. 223.
49 Ibid.

A study of cosmic redemption has been more recent. Johnson begins by revisiting scriptural texts that are inclusive of the whole of creation, for example, the hymn in Colossians that celebrates Jesus, in whom 'the fullness of God was pleased to dwell, and through him God was pleased to reconcile to himself all things, whether on earth or in heaven, making peace through the blood of his cross' (Col 1:20). The book of Revelation envisions the end time, when all things will be made new (Rev 21:5). Johnson acknowledges the more cosmic scope of theologies of redemption in the Eastern Orthodox tradition, before taking up Rahner's discussion of the cross. Rahner highlights God's gift of grace in creation and the offer of forgiveness through the incarnation. Liberation comes from the realisation of God's love and forgiveness, rather than from the sacrificial offering of Jesus on the cross. Rahner was influenced by a school of thought coming from the Franciscan Duns Scotus (c. 1266–1308), who, in the words of Johnson, 'maintained that the incarnation would have taken place whether human beings had sinned or not.'[50] This position distinguished Scotus from Aquinas and others who argued that 'the world was created good; our first parents sinned, ruining their relationship with God; therefore, in mercy, the Son of God became incarnate and died in order to restore the relationship'[51]. Scotus' position, in contrast, is about the dynamic of God's unfathomable love, which is always actively seeking union. The death of Jesus is still about redemption, but Johnson interprets its efficacy here, as in her earlier work, 'not in satisfaction rendered to a God whose honor has been violated, but in the presence of divine love in the flesh enacting an historical solidarity with all who suffer and die.'[52]

The cosmic community is not capable of sinning, but it is blatantly finite. Its transfiguration is part of resurrection hope. The bodily Jesus rose into a new creation, one that foreshadows the future for all reality. Johnson sums up by saying, 'Christ carries the whole creation towards it destiny. His resurrection is the beginning of the resurrection of all flesh. Or so Christians hope.'[53] This is the language of faith, the faith

---

50 Ibid., p. 226.
51 Ibid.
52 Ibid.
53 Ibid., p. 227.

that 'the encompassing mystery enacted in Jesus Christ through the Spirit bears creation forward with an unimaginable promise toward a final fulfilment when God will be "all in all" (1 Cor 15:28).'[54]

Johnson concludes her reflection on this movement in hope and faith into fullness of life in God with a consideration of animal heaven. Those with pets and those who advocate for compassionate treatment of animals and for the health of ecosystems are concerned about this issue. Traditional perspectives exclude other-than-human species from a future of wholeness and fullness of life in God. Aquinas, for example, as presented by Johnson, sees plants and animals as valuable only for their usefulness to humans; subsequently, when humans cease to need them, they will have no future. Those who see the symmetry between creation and redemption recognise the biophysical world as created and valuable in its own right, and for them Johnson's position is feasible:

> Affirming that the promise of new creation includes all creatures as individuals in a way appropriate to their nature is not a foolish construal. Based on the belief that the Giver of life indwells each creature to empower its life within the evolutionary process, and that the same Spirit of the crucified and risen Christ accompanies each creature in its pain and dying, this position figures it would be discordant with the fibre of creative love to allow any creature just to disappear.[55]

Up to this point, Johnsons' focus has been on the 'theological meaning of the natural world of life'.[56] She has not allowed the intensity of her gaze to veer towards an anthropocentric view, according to which the natural world could derive its meaning only from its human connection. In this she really attempts to listen to the beasts:

> Continuously fired into being by the Giver of life, the living world is the dwelling place of God. Ontologically dependent on the Creator, it is empowered with the autonomy befitting a finite creature to operate freely in the course of its own evolution. In solidarity with the perishing of Christ who shares its flesh, it is a groaning, cruciform

---

54 Ibid.
55 Ibid., p. 233.
56 Ibid., p. xv.

world, destined for resurrection. Existing in absolute dependence on its Maker, it bears the promise of new eschatological life, heading toward a final fulfilment, thanks to the *Alpha* and *Omega* whose fidelity knows no end.[57]

In Principles 3 and 4, Earth Link draws on Johnson's Trinitarian theology of the Spirit, and contemporary Christology, creation theology and eschatology as a basis for Christian ecospirituality. From this theology, Earth Link derives important perspectives on Earth as God's dwelling place, and on Jesus as the God of love embracing the cosmos in healing solidarity. With the eyes of faith we can say that our encounter with cosmos and planet can indeed be encounters with God present, revealing, healing, and opening up a future full of hope for the whole cosmos.

It is now time to look more specifically at humans within the Earth community and, indeed, within the community of creation, and the ethical responsibilities that flow from this investigation. This is the subject of Principle 5.

---

57 Ibid., p. 235.

.

CHAPTER 8

Principle 5: The Basis of Right Relations

Live in right relations within the community of creation.

Principle 5 follows on from the preceding principles, which address the contextual imperative; the open, attentive and receptive attitudes that orient one to the ecospiritual experience; and the Christian contribution that arises out of a belief in God's continuing and creative presence and promise, and out of Jesus Christ's assuming a cosmic identity.

The original Principle 5 read: 'Live in right relationships within the interdependent web of life.' It stemmed from 'a belief in the intrinsic value of all species, and from deep bonding with the earth.' This principle was recognised as 'the culmination of the spiritual dynamic' established in the preceding principles. We then explored the nature of 'right relationships'. All this is carried forward in the enhanced principles. Enhanced Principle 5 is about right relations within the community of creation. This principle challenges the world view of hierarchical dualism, situates humans within the evolutionary process, and locates the whole Earth community within the embrace of God, Ground of all being, who is drawing everything forth to fullness of life. The explication of this principle draws on the work of Elizabeth Johnson, supplemented by that of other writers.

Humans are within the community of life. Our materiality continuously derives from the stuff of stardust, and 'we humans ... share a genetic heritage with every other species of the tree of life, a biological kinship encoded in each cell of our body.'[1] Our distinguishing capacities of 'self-reflective consciousness and freedom,'[2] or of 'mind and will,' to use the classical terms, were new capacities for the universe. This is

---

1 Johnson, *Ask the Beasts*, p. 237.
2 Ibid., p. 236.

the evolutionary world view espoused in these principles, and it is here explored for its theological significance.

Humans within evolution

Like Earth Link, Johnson locates our species, *homo sapiens,* within the story of evolution. Our story is out of Africa, with the emergence of primates, which diversified into gorilla, chimpanzee and hominid lines.[3] Our immediate hominin ancestors, *Australopithecus,* were bipedal and moved increasingly to live in the open savannah rather than in the forests. The earliest expression of the genus homo was *homo habilis,* and according to the fossil records, it and its successors had physical capacities and cultural abilities that indicate the development of mental processes. This was not a linear development in any one geographical location. Variations on the theme are being found in various parts of Africa, Europe and Asia, and many of the lines died out to the point where *homo sapiens* is now the only survivor.

Johnson points out the distinguishing characteristics of *homo sapiens* as 'self-consciousness, use of language, and tremendous fluidity in behaviour.'[4] Given the importance of establishing the evolutionary emergence of mind, I am also drawing on Bellah's work, in which he draws parallels between the developmental stages of children those of human culture. Bellah identifies that children pass through stages, which he calls unitive, enactive, symbolic, and conceptual representation.[5] Human culture passes through corresponding stages in its movement through mammalian attentiveness, early human capacities for mime and gesture, the later development of mythic or symbolic thought and a capacity for ritual, to archaic society round 1,000 BCE with its capacity for conceptual and theoretical thought. It is at that stage that Bellah locates the emerging capacity for religion arising in association with concepts of kingship and divinity.

---

3 Ibid., p. 237. Note that 'many of our extinct ancestors are now called hominins. However, it's not technically wrong to call them hominids – all members of Hominini are also members of the subfamily Homininae and the family Hominidae.' https://australianmuseum.net.au/learn/science/human-evolution/hominid-and-hominin-whats-the-difference/.

4 Johnson, *Ask the Beasts*, p. 239.

5 Bellah, *Religion in Human Evolution*, p. 117 .

Johnson in turn marvels at the 'intellectual and volitional powers'[6] of the mind, which science explains in terms of its material functions, but which can be seen as evidence of the emergence of new capacities requiring new levels of explanation. She points out that the field of understanding mind-brain relationships is currently a very lively area of research.

Johnson goes on to highlight the distinctiveness of human consciousness, which by now is a player in the very process of evolution. She lists a number of ways in which the human is described:

> In view of their singular self-reflective inwardness, cognitive powers, and freedom of action, their philosophers describe them as persons composed of body and soul, rational animals, spirited selves, embodied spirits, spirit in the world. Religious teachers add that they are created in the divine image and likeness, being a complex unity whose body comes from the dust of the earth, and whose soul is breathed into them by God, each one gifted with unique dignity.[7]

Johnson moves on to address humans in relationship with the community of life before retrieving a Christian anthropology to build on her sense of human embeddedness in the evolution of life on Earth. Similarly, Earth Link is concerned about its place within the Earth community and, indeed, the community of creation.

Negative human impacts on the community of life

Johnson moves on to ethical considerations. The human species is currently having a very mixed impact on the Earth community of which it is part. The richness of Earth's biodiversity is threatened. Some of this has been addressed under Principle 1. Now is the moment for facing and addressing negative human impacts. Johnson draws on McKibben's recent work *Eaarth*[8] and other sources to paint a picture of the geophysical effect that humans are now having on this planet. She draws attention specifically to population growth, resource consumption, pollution and

---

6 Johnson, *Ask the Beasts,* p. 239.

7 Ibid., p. 240.

8 McKibben, *Eaarth.*

the extinction of species, all of which document the reality of an Earth in crisis, and concludes that

> [i]f human beings were to wake up to the grandeur of the dying world, fall in love with life, and change their behaviour to protect it, much of the current dying off could be slowly brought under control. But in our day the dire situation appears to be accelerating, with humanity's rapacious habits driving species to extinction faster than new species are able to evolve. The tree of life is thinning out.[9]

Johnson moves from the scenario of destruction to a call to conversion, in much the same way as *Laudato Si* moves from its statement of the problems, which draws on the best knowledge available, to a vision of integral ecology and a call to action at every level of society. There are good reasons for attending to the state of the planet. One is intergenerational in the sense of maintaining the planet for future generations. Another is Haught's sense of the 'promissory character of the natural world ... [which] is due to the inexhaustible vitality of God who created it.'[10] We need our evolving world to manifest the potential for increasing complexity and vibrancy. It is full of promise that we cannot abort by our carelessness. Johnson adds the theological perspective, which should urge us to greater responsibility:

> In its continuous creation by the empowering spirit, its redeeming solidarity in the flesh of the crucified and risen Jesus Christ, its origin and ultimate future in the faithful love of the creator, and its sacramental and revelatory character in all concrete beauty, suffering and surprise – from every theological angle the tree of life calls forth respect and responsible love. As its bare, natural, evolving self, it is worthy of this.[11]

The failure to respond in this way is being called out by ethicists as the sins of biocide, ecocide, or geocide. Such devastation of the planet has repercussions at many levels. Johnson quotes the Catholic Bishops of the Philippines, who see that such treatment of nature 'defaces the

---

9 Johnson, *Ask the Beasts,* p. 253.
10 Ibid., p. 254.
11 Ibid., p. 255.

image of Christ which is etched in creation.'[12] The Catholic tradition has taken the lead since Pope John Paul II's call for ecological conversion in his 1990 message for the World Day of Peace. It continues in the recent encyclical of Pope Francis, who draws on the Eastern Orthodox tradition in also classifying the abuse of nature as sin – sin against Earth and her people:

> For human beings ... to destroy the biological diversity of God's creation; for human beings to degrade the integrity of the earth by causing changes in its climate, by stripping the earth of its natural forests or destroying its wetlands; for human beings to contaminate the earth's waters, its land, its air, and its life – these are sins.[13]

Johnson lauds the extension in Catholic thought of respect for life as embracing both human life and the rest of creation. The moral imperative extends to both. Our failing in this area requires repentance and a change of heart. Johnson spells out the intellectual, emotional and ethical dimensions of such a conversion, and says quite lyrically, 'In sum, ecological conversion means falling in love with earth as an inherently valuable, living community in which we participate, and bending every effort to be creatively faithful to its well-being, in tune with the living God who brought it into being and cherishes it with unconditional love.'[14]

In a similar vein, Pope Francis uses emotive language to capture the felt conversion that is needed to foster a spirit of generous care, full of tenderness:

> First, it entails gratitude and gratuitousness, a recognition that the world is God's loving gift, and that we are called quietly to imitate his generosity in self-sacrifice and good works: '[D]o not let your left hand know what your right hand is doing ... and your Father who sees in secret will reward you' (cf. Mt 6:3–4). It also entails a loving awareness that we are not disconnected from the rest of creatures, but joined in a splendid universal communion. As believers, we do not look at the world from without but from within, conscious

---

12 Catholic Bishops of the Philippines, 'What is Happening?', p. 316.
13 Address in Santa Barbara, California, 8 November 1997; cf. Chryssavgis, *On Earth.*
14 Johnson, *Ask the Beasts,* p. 259.

of the bonds with which the Father has linked us with all beings. By developing our individual, God-given capacities, an ecological conversion can inspire us to greater creativity and enthusiasm in resolving the world's problems and in offering ourselves to God 'as a living sacrifice, holy and acceptable' (Rom 12:1).[15]

So why is this not happening on a larger scale? Johnson's book culminates with her vision of a new paradigm to replace the current overarching paradigm of dominion, which, directly or indirectly, can be seen as the root of oppression of nature. In like manner, *Laudato Si* presents a vision that is very much about communion with nature, with other humans and with God. This is the vision embraced by Earth Link with its commitment to 'respect, reverence and care for the whole Earth community, held as it is in the embrace of the Divine' and its mission to facilitate 'deep bonding within the community of creation.' But first to Johnson, to establish her vision as viable and important in the pursuit of the Earth Link vision and mission.

A new paradigm – from dominion to communion

Something is not working if we have moved to the current state of environmental degradation. Something is not working if we even need an initiative such as Earth Link. In the course of asking the beasts, Johnson has found treasures in the tradition that indicate that 'loving life on earth, far from being foreign to the living tradition of Christianity, is actually supported by its core cherished beliefs about God revealed in scripture and condensed in the creed.'[16] There is promise in the tradition and there are problems. There is a history of inherited oppression. In her earlier works, Johnson pointed to the world view of hierarchical dualism, embedded as it is in religion and culture, as leading to a devaluing of nature. In *Ask the Beasts* she points to the tradition of human dominion embedded in the creation narrative in Genesis 1:28. It paints 'human beings at the apex of the pyramid of living creatures with rights over otherkind.'[17] Johnson puts a more positive spin on the problematic text.

---

15 Pope Francis, *Laudato Si*, p. 220.
16 Johnson, *Ask the Beasts,* p. 260.
17 Ibid., p. 261.

The courtly origins of the tradition that gave rise to the first Genesis narrative point to a positive interpretation that sees the text as being about exercising responsibility delegated by the ruler. The mandate to preserve species is there in the story of Noah and the ark in Genesis 6:19. The creation story in the second chapter of Genesis has quite a different tone from that of the first chapter of Genesis. Adam, the Earth creature, is given the mandate to 'till [the earth] and keep it'(Gen 2:5) which is generally interpreted as about cultivation and care. This narrative is seen as the basis of kinship, of 'the earthly solidarity women and men have with each other and the rest of creation.'[18] Cardinal Turkson, in a speech given not long before the launch of *Laudato Si*, suggested that humans had been doing too much tilling and not enough keeping![19] The attitude he is critiquing would be reinforced by the problematic text of Psalm 8:6, which speaks this way of God's design: 'You have given them dominion over the works of your hands; you have put all things under their feet.'

Laudato Si is strong in its insistence that Genesis 1:28 be interpreted more appropriately. Even if 'we Christians have at times incorrectly interpreted the Scriptures, nowadays we must forcefully reject the notion that our being created in God's image and given dominion over the earth justifies absolutes domination over other creatures.'[20] Instead, 'we are called to recognize that other living beings have a value of their own in God's eyes … In our time the Church does not simply state that other creatures are completely subordinate to the good of human beings, as if they had no worth in themselves and can be treated as we wish.'[21] Rather, '[T]hey have an intrinsic value independent of their usefulness,'[22] because God loves them. 'Even the fleeting life of the least of beings is the object of God's love, and in its few seconds of existence, God enfolds it with affection.'[23]

But damage has been done. The hierarchical world view so typical of the priestly and royal systems of the time has passed over into western

---

18 Ibid., p. 264.
19 Turkson, 'Integral Ecology and the Horizon of Hope'.
20 Pope Francis, *Laudato Si*, p. 67.
21 Ibid.
22 Ibid., p. 140.
23 Ibid., p. 77.

culture. As noted under Principle 1, it underpins Western economic, social and environmental norms to the present day. As Johnson remarks, 'Slipping its biblical bonds, the notion of dominion has supported rampant use and abuse of the earth.'[24] At its best, the call to responsible stewardship has curbed some excesses, and its core tenets are laudable. 'A steward is a person who manages another's property or financial affairs, one who administers material wealth as the agent of another. The core of theological stewardship is the belief that the earth and all of its resources belong ultimately to God. With overwhelming generosity, God entrusts these good things to human beings, gifting us with their use.'[25]

Increasingly, problems are being identified with the notion of stewardship within a universe that is emerging. Johnson points to a few of these. One is that '[i]t envisions humans being independent from the rest of creation and external to its functioning. Lacking a deep ecological sensibility, it establishes a vertical top-down relationship, giving human beings responsible mastery over other creatures but not alongside them or open to their giving.'[26] Such a model is not adequate to underpin the conversion to responsible use of resources, especially for those who have developed a world view based on evolutionary emergence. Johnson suggests that there needs to be 'a different conceptuality of the human place in the world, religiously speaking. Such an alternative presents itself in the biblical view of the community of creation.'[27]

Pope Francis goes some of the way towards this conclusion when he acknowledges the problematic issue of a culture based on human mastery over nature, which he refers to as 'excessive anthropocentrism':

> Modernity has been marked by an excessive anthropocentrism which today, under another guise, continues to stand in the way of shared understanding and of any effort to strengthen social bonds. The time has come to pay renewed attention to reality and the limits it imposes; these in turn are the condition for a more sound and fruitful development of individuals and society. An inadequate presentation

---

24 Johnson, *Ask the Beasts,* p. 265.
25 Ibid.
26 Ibid., p. 266.
27 Ibid., p. 267.

of Christian anthropology gave rise to a wrong understanding of the relationship between human beings and the world. Often, what was handed on was a Promethean vision of mastery over the world, which gave the impression that the protection of nature was something only the faint-hearted cared about. Instead, our 'dominion' over the universe should be understood more properly in the sense of responsible stewardship.[28]

Pope Francis does not reject outright the notion of anthropocentrism but critiques its excesses. He warns against replacing it with biocentrism, with its potential for denying that human beings possess 'a particular dignity above other creatures,'[29] and insists that '[A] correct relationship with the created world demands that we not weaken this social dimension of openness to others, and much less the transcendent dimension of our openness to the 'Thou' of God.'[30] I would argue that the dignity of human persons can be respected by acknowledging their distinctive place based on their unique capacities, without resorting to hierarchical language. There is an intermediate position between anthropocentrism and biocentrism that locates humans in relationship with nature and the Divine without any claim to equality and with full recognition of their uniqueness. Johnson's paradigm of the community of life and the community of creation contributes much to this understanding.

Johnson develops the biblical vision of the *community of creation*. Evolutionary biology establishes the existence of a community of life, also referred to as the Earth community:

> Historically, all life results from the same biological process; genetically, living beings share elements of the same basic code; functionally species interact without ceasing. Human beings belong to this community and need other species profoundly, in ways more than other species need them.[31]

Rather than being apart from the community of life and the cosmos, humans are within them, a world view previously referred to in Principle

---

28 Pope Francis, *Laudato Si*, p. 116.
29 Ibid., p. 119.
30 Ibid.
31 Johnson, *Ask the Beasts*, p. 267.

1 as anthropocosmic. Enter a theocentric world view that orients such an interdependent world to God. As established earlier, when we explored the religious significance of this interdependence, the whole world with all its members 'in its origin, history and goal … is ultimately grounded in the creative, redeeming God of love.' The theological construct of the community of creation is founded on the belief that 'all beings are in fact creatures, sustained in life by the Creator of all that is.' This applies to humankind and other species, and this commonality before God is stronger than their differences. In their kinship all are 'grounded in absolute, universal reliance on the living God for the breath of life.'[32] This pattern of relationship, which locates us humans alongside other creatures and stresses interconnectedness without blurring differences, gives a new impetus for ethical behaviour based on relationships with one another and the wider whole, a perspective that can supersede notions of dominion. This is the world view espoused by Earth Link.

Biblically this notion is rich. In *Ask the Beasts,* Johnson looked at the community of life through the eyes of Job. She now embarks on an extended study of Job chapters 38–41, and her findings are very relevant to this principle, so I synthesise them in some detail. Job has come upon hard times, and his friends argue that he must have sinned. Job maintains his innocence, but his suffering is very real. Eventually God speaks to him out of the whirlwind, and asks, 'Where were you when I laid the foundation of the earth?' (Job 38:4). The questions go on and on as God draws attention to the scope of the physical world, and the qualities of the wild yet free animals and the fearsome beasts. These creations are a far cry from anything that is subject to human dominion, yet they reveal their Creator.

Johnson draws out several points from commentaries on the book of Job. It is preoccupied with the meaning of suffering, in this case the suffering of a good person. It concludes is that, ultimately, suffering is a mystery to be respected. God rebukes Job's friends, who want to insist that Job's suffering is sent from God as punishment for sin. Despite the focus on suffering, these chapters are brimming with wonder and power,

---

32 Ibid., p. 268. This page in Johnson's work is the source of all ensuing quotes up to the next footnote.

and Job shifts his position, saying 'I had heard of you by the hearing of the ear, but now my eye sees you' (Job 42:5). Johnson sees this passage as modelling a relationship of humans with God and the rest of creation. Far from seeing human dominion, 'Job is led to see divine activity in the awesome, independent working of the natural world over which he has no mastery, not only technologically but also theologically.'[33] The creatures have their own value, and in observing them, Job finds wisdom and a recognition of his own place within what Johnson calls the community of creation. Our assumptions of human superiority are seriously challenged by this narrative. There can be a very positive outcome if 'humbled and delighted by the other life around us, we can grow to know ourselves as members of the community of creation and step up to protect our kin.'[34] In fact, this is at the heart of ecological conversion.

Johnson continues to make a case for the community of creation, and for the place of humans within that community, by referring to that great creation song of praise, Psalm 104, to which I have referred earlier in this book. Once again it is by observing the whole community of life, the sky, the earth, the sea, the vegetation, the animals, the people and the rhythms of the day, that we discern God's presence and give praise.

In a theocentric world view, people take their place 'within the wider world which enjoys its own direct relation to the Creator.'[35] In Psalm 148, old and young people take their place with angels, sun, moon, mountains, hills, trees, princes and rulers in actively giving praise to the Creator who gave them life. Johnson makes a point about the order in which God's creation is named in this psalm. While the humans may come at the end, the angels come first, and she notes that

> [r]ather than the pattern of dominion which climaxes the appearance of humans in Genesis 1, and different from the hierarchical chain of being found in medieval thought, here is an interwoven assembly of everything from sky, sea, and land, each one part of a grateful community of creation praising God.[36]

---

33 Ibid., p. 272.
34 Ibid., p. 273.
35 Ibid., p. 275.
36 Ibid., p. 276.

Together they praise God without consideration of who is the most valuable. Johnson then addresses whether all creatures are capable of giving such praise. She makes an important point when she notes that such a metaphor is an extension of the human capacity to give praise, but one that contains a valuable insight. 'By virtue of their being created, of being held in existence by the loving power of the Creator Spirit, all beings give glory to God simply by being themselves.'[37] By their very existence they are oriented to God, as we are. Our challenge is to be attentive to this and join in the song that they are already singing. This can remind us of our place in the community of creation. 'At this time of ecological catastrophe, praying with a sense of participation in creation's praise of God allows people to recover a healthy sense of their own human place in the world as created beings alongside our fellow creatures.'[38]

Moving from the Psalms to the prophets, Johnson listens to Creation mourning and lamenting, often as a result of human action. The fate of the people and of Earth are interconnected. Hosea laments that

[t]here is no faithfulness or loyalty, and no knowledge
of God in the land.
Swearing, lying, and murder, and stealing and adultery break out;
bloodshed follows bloodshed.
Therefore the land mourns, and all who live in it languish;
together with the wild animals and the birds of the air,
even the fish of the sea are perishing (Hos 4:1b–3).

As with the experience of land giving praise, we have the image of land mourning. This is devastation caused by human action, which ought to be a familiar theme to us. Negative actions by one part of the community of life have implications for the rest of the community. All is not lost, however, and just as Paul in Romans 8:18–23 hears creation groaning in the hope of being set free, so the prophets proclaim 'that the mercy and steadfast love of God will establish justice in a disordered world, and this hope is announced with a vigor equal to their

---

37 Ibid.
38 Ibid., p. 278.

denunciation of human wrong-doing.'[39]

Johnson sums up this vision of the community of creation and extols its potential in words that locate humans in relation to God and to the rest of the community of creation. It is appropriate to quote this fully, as it provides is a key insight for the work of Earth Link's commitment to facilitate moving to a new paradigm:

> The biblical vision of the community of creation opens a life enhancing avenue of relationship. Departing from a long history of interpretation, it scoops up the Genesis notion of dominion and places it within the mutual interactions of all beings as creatures in relation to the living God who creates and redeems. The community model brings forward at the most fundamental level our theological human identity as created, our biological embeddedness in the natural world, and our reciprocal interdependence with the other species and the life-giving systems that support us all.[40]

This is the context in which we live out what she calls the ecological vocation. Our call to love extends to the whole community of creation. Pope Francis calls this the call to care for our common home. Johnson acknowledges the adoption of various religious practices such as contemplation, asceticism, and prophetic action for justice, reframed in ecological terms. She admits that 'the prophetic dimension of the ecological vocation still beckons the churches for the most part.'[41] It is not yet a reality.

For the Catholic Church and beyond, *Laudato Si* is a very important response, and its favourable reception in so many quarters has demonstrated that the churches have a role to play that can be welcomed by the broader society. It presents a vision in which communion is paramount. The word features at least a dozen times in the encyclical. At other times, there is a refrain that 'everything is connected.'[42] At times interconnectedness is the vision, at times the diagnosis, and always part of the solution.

---

39 Ibid., p. 280.
40 Ibid.
41 Ibid., p. 284.
42 Pope Francis, *Laudato Si*, p. 91.

This encyclical acknowledges, in a way that is new to formal documents of the Church, that 'because all creatures are connected, each must be cherished with love and respect, for all living creatures are dependent on one another.'[43] Humans are located within this community. 'Nature cannot be regarded as something separate from ourselves or as a mere setting in which we live. We are part of nature, included in it and thus in constant interaction with it.'[44]

This interconnectedness is stressed in other parts of *Laudato Si* as well:

> This is the basis of our conviction that, as part of the universe, called into being by one Father, all of us are linked by unseen bonds and together form a kind of universal family, a sublime communion which fills us with a sacred, affectionate and humble respect. Here I would reiterate that God has joined us so closely to the world around us that we can feel the desertification of the soil almost as a physical ailment, and the extinction of a species as a painful disfigurement.[45]

Our creation draws us into communion. Christ who 'incorporated into in his person part of the material world'[46] draws us into God's fullness. Through our immersion in this web of relationships, we are reminded that 'everything is interconnected, and this invites us to develop a spirituality of that global solidarity which flows from the mystery of the Trinity.'[47] The challenges are great, but so are the sources of hope:

> In the heart of this world, the Lord of life, who loves us so much, is always present. He does not abandon us, he does not leave us alone, for he has united himself definitively to our earth, and his love constantly impels us to find new ways forward. Praise be to him![48]

In 2016, after the publication of *Ask the Beasts*, and after the encyclical had been promulgated, Johnson identified 'one theme that runs like a

---

43 Ibid., p. 42.
44 Ibid., p. 139.
45 Ibid., p. 89.
46 Ibid., p. 235.
47 Ibid., p. 240.
48 Ibid., p. 245. The use of masculine pronouns is, unfortunately, in the official English translation.

silver thread through all of the encyclical's multifaceted teaching. This is the sacred value of living plants and animals in and of themselves in a community to which human beings also belong.'[49] She clearly identifies the consonance between *Ask the Beasts* and *Laudato Si.*

In an exhortatory style similar to the final paragraph of the encyclical quoted above, Johnson concludes *Ask the Beasts* by recapping her movement from asking the beasts what they could teach us, to listening to their scientific, theological and ecological messages. She reminds us of the urgency of taking up a commitment to ecological wholeness if we are to move towards the vision that must guide us at this critical time, namely, that of 'a flourishing humanity on a thriving planet, rich in species in an evolving universe, all together filled with the glory of God.'[50] Earth Link fully embraces this vision.

In conclusion

The enhanced principles of Earth Link have the potential to facilitate movement toward the implementation of its vision and mission. They also have the potential to show how 'loving the Earth arises as an intrinsic part of faith in God, rather than just an add-on.'[51] They look outward to the context that necessitates the call to ecological conversion in order to address the reality of environmental degradation. With Thomas Berry, they consider what is at the heart of the ecospiritual experience, namely, an openness and receptivity to the presence of the Spirit in created reality. With Elizabeth Johnson and Pope Francis, they examine theological understandings that enable us to name the ecospiritual encounter as one with God dwelling in cosmos and Earth in a way that does not deny its dynamism in the process of evolutionary emergence, nor contain God within our finite reality. These principles locate Earth Link firmly within the Judaeo-Christian tradition by acknowledging Jesus Christ as the God of Love who embraced the cosmos in healing solidarity, and they call us humans into right relations with God, self, others and all of creation. These principles can nourish us even as they challenge us to a

49 Johnson, 'From Pyramid to Circle', p. 486.

50 Johnson, *Ask the Beasts,* p. 286.

51 Johnson, 'From Pyramid to Circle', p. 479.

future that has to be different. They call us to love Earth, to listen to it, to 'respect, reverence and care for the whole Earth community,'[52] and in so doing to encounter the God of love who is present now, holding out a promise of wholeness and fullness of life.

The enhanced Principles of Earth Link elaborate a world view of a dynamic and unfolding cosmos, an amplified approach to ecospirituality, and the inclusion of a more specifically Christian theological framework. Earth Link had intended taking the original principles into dialogue with various faith traditions. Earth Link is grounded in the Christian traditions of Catholicism and the Sisters of Mercy. This is our field of belonging and influence. For those for whom this tradition is meaningful, the ecospiritual encounter needs to be able to be articulated in a way that is more specific than simply an encounter with the Sacred. Based on their historical experiences, traditions name the Sacred in various ways. The enhanced principles offer a framework and language that emerge from the Christian tradition. They recognise some of its problems and explore its potential to contribute to urgent ecological concerns.

Through its enhanced principles, Earth Link and its community are the immediate beneficiaries of this exercise of practical theology on the theme of the embrace of Heaven and Earth, and the way in which ecospirituality in an emerging universe can contribute to living out such a vision. The benefits can ripple out from there. As indicated in the introduction to this book, this research is necessary to enrich the integration of faith and life for those who want a better world for themselves and their children. It provides a vital underpinning for their individual and collective ways of living sustainably, as an approach to work and ministry, and to action for ecojustice. For those who are committed, it can show what living out a new paradigm can look like. For those who are dubious, it has the potential to underline the vital connection between faith and ecological responsibility. For those who are critical, these principles demonstrate the deeply theological basis of ecospiritual practice and concern for the environment.

This research can make a practical contribution by linking

52 From the vision statement of Earth Link, www.earth-link.org.au

understanding and action. Its hoped-for outcome for the Earth Link community and beyond is ecological conversion, understood as 'falling in love with earth as an inherently valuable, living community in which we participate, and bending every effort to be creatively faithful to its well-being, in tune with the living God who brought it into being and cherishes it with unconditional love.'[53]

This book has established that the way people of faith envision the embrace of Heaven and Earth determines how they relate to the Earth community and, indeed, to the whole community of creation. For the most part, one does not abuse what one holds in a loving embrace. The enhanced principles celebrate not only the embrace of Heaven and Earth, but the mystery proclaimed in the *Exultet*, that song of joy used during the Easter vigil. It speaks of the 'truly blessed night, when things of heaven are wed to those of earth, and divine to human.' The embrace becomes a wedding. We are part of one of the greatest love stories that can ever be told, and that behoves us to live in right relations with God and with all that is.

---

53 Johnson, *Ask the Beasts,* p. 259.

Bibliography

Abrams, Nancy Ellen, & Primack, Joel R., *The New Universe and the Human Future*, Yale University Press, New Haven, CT, 2011.

———, *The View from the Centre of the Universe*, Fourth Estate, London, 2006.

Augustine. *Sermons III/3 (The Works of Saint Augustine: A Translation for the 21ˢᵗ Century)*, translated by Edmund Hill O.P., New City Press, Brooklyn, NY, 1991, pp. 225–6.

Baker, Susan & Morrison, Robin. 'Environmental Spirituality: Grounding our Response to Climate Change', *European Journal of Science and Theology*, vol. 4, no. 2, 2008, pp. 35–50.

Barbour, Ian. *Nature, Human Nature and God*, Fortress Press, Minneapolis, MN, 2002.

———, 'Response: Ian Barbour on Typologies', *Zygon*, vol. 37, no. 2, 2002, pp. 345–59.

Bellah, Robert N. *Religion in Human Evolution: From the Palaeolithic to the Axial Age.* Belknap Press of Harvard University Press, Cambridge, MA, 2011.

Berry, Thomas. *Christian Future and the Fate of the Earth,* edited by Mary Evelyn Tucker and John Grim, Orbis Books, London, 2009.

———, 'The Emerging Ecozoic Period', in Ervin Laslo and Allan Combs (eds), *Thomas Berry, Dreamer of the Earth*, Inner Tradition, Rochester, VT, 2011, pp. 6–15.

———, *Evening Thoughts: Reflecting on Earth as a Sacred Community,* edited by Mary Evelyn Tucker, Sierra Club Books, San Franscisco, CA,, 2006.

———, *The Great Work: Our Way into the Future*, Bell Tower/Random House, New York, 1998.

———, *The Sacred Universe: Earth, Spirituality and Religion in the*

Twenty-First Century, edited by Mary Evelyn Tucker, Columbia University Press, New York, 2009.

———, 'The Spirituality of the Earth', in Charles Birch, William Eakin and Jay B. McDaniel (eds), *Liberating Life: Contemporary Approaches in Ecotheological Theology*, Orbis Books, Maryknoll, New York, 1990, pp. 151–8.

Bird-David, Nurit. "Animism" Revisited: Personhood, Environment and Relational Epistemology', *Current Anthropology*, vol. 40, no. S1, 1999, pp. 67–91.

Bridle, Susan. 'Comprehensive Compassion: An Interview with Brian Swimme', in *What is Enlightenment?,* issue 19, 2 July 2003, http://thegreatstory.org/SwimmeWIE.pdf.

Browning, Don S. *A Fundamental Practical Theology*, Fortress Press, Minneapolis,MN, 1991.

Catholic Bishops of the Philippines, 'What is Happening to our Beautiful Land? A Pastoral Letter on Ecology', in Drew Christensen and Walter Grazer (eds), *And God Saw that It Was Good: Catholic Theology and the Environment*, United States Conference of Catholic Bishops, Washington, DC, 1996.

Clarke, W. Norris. *Explorations in Metaphysics: Being-God-Person,* University of Notre Dame, Notre Dame, IN, 1994.

Clayton, Philip. 'Toward a Constructive Christian Theology of Emergence', in *Evolution and Emergence: Systems, Organisms, Persons,* Oxford University Press Oxford, 2007, http://site.ebrary.com/id/10271643?ppg=330.

Combs, Allan. *The Radiance of Being: Understanding the Grand Integral Vision, Living the Integral Life,* Paragon House, St Paul, MN, 2002.

Committee on Doctrine, United States Conference of Catholic Bishops (USCCB). 'Statement on Quest for the Living God: Mapping Frontiers in the Theology of God by Sister Elizabeth Johnson.', in Richard R. Galliardetz (ed.), *When The Magisterium Intervenes: The Magisterium and Theologians in Today's Church*, Liturgical Press, Collegeville, MN, 2012, pp. 180–89. .

————, in Richard R. Gaillardetz (ed.), *When the Magisterium Intervenes: The Magisterium and Theologians in Today's Church*, Liturgical Press, Collegeville, MN, 2012, pp. 256–60.

Costigan, Philip, Rose, Patricia and Tinney, Mary. 'Introduction to Ecospirituality', Earth Link, 2007.

Costigan, Philip, Rose, Patricia and Tinney, Mary, 'Role of Spirituality', *Social Alternatives,* vol. 26, no. 3, Third Quarter, 2007.

Crowley, Paul. 'Book Discussion: Elizabeth Johnson's Ask the Beasts: Darwin and the God of Love', *Theological Studies* vol. 77, no. 2, 2016, pp. 466–87.

Darragh, Neil. 'The Practice of Practical Theology: Key Decisions and Abiding Hazards in Doing Practical Theology', *Australian eJournal of Practical Theology*, vol. 9, 2007, pp. 1–12, http://aejt.com.au/__data/assets/pdf_file/0006/395736/AEJT_9.9_Darragh_Practice.pdf.

Darwin, Charles. *The Annotated Origin: A Facsimile of the First Edition of On the Origin of Species,* Belknap Press of Harvard University Press, Cambridge, MA, 2011.

Deacon, Terrence W, & Goodenough, Ursula. 'The Sacred Emergence of Nature', in *The Oxford Handbook of Religion and Science*, Oxford University Press, New York, 2006, pp. 853–71.

Delio, Ilia. *Christ in Evolution,* Orbis Books, Maryknoll, NY, 2008.

————, *The Emergent Christ: Exploring the Meaning of Catholic in an Evolutionary Universe*, Orbis Books, Maryknoll, NY, 2011.

————, *The Unbearable Wholeness of Being: God, Evolution, and the Power of Love*, Orbis Books, Maryknoll, N Y, 2013.

Donald, Merlin. *Origin of the Modern Mind: Three Stages in the Evolution of Culture and Cognition*, Harvard University Press, Cambridge, MA, 1991.

Eaton, Heather. 'Metamorphosis: A Cosmology of Religions in an Ecological Age', in *The Intellectual Journey of Thomas Berry: Imagining the Earth Community*, Lexington Books, Boston, MA, 2014, pp. 149–71.

Ecumenical Patriarch Bartholomew. *On Earth as in Heaven: Ecological Vision and Initiatives of Ecumenical Patriarch Bartholomew*, edited by John Chryssavgis, Fordham University Press, New York, 2012.

Edwards, Denis. 'Christology in the Meeting between Science and Religion: A Tribute to Ian Barbour', *Theology and Science*, vol. 3, no. 2, 2005, pp. 211–20, https://doi.org/10.1080/14746700500141747.

———, *Partaking of God, Trinity, Evolution and Ecology*, Liturgical Press, Collegeville, MN, 2014.

———, 'Sublime Communion: The Theology of the Natural World in Laudato Si', *Theological Studies*, vol. 77, no. 2, 2016, pp. 377–91.

Fox, Patricia. *God as Communion: John Zizioulas, Elizabeth Johnson, and the Retrieval of the Symbol of the Triune God*, Liturgical Press, Collegeville, MN, 2001.

Francis, *Laudato Si: On Care for Our Common Home*, 2015, http://w2.vatican.va/content/francesco/en/encyclicals/documents/papa-francesco_20150524_enciclica-laudato-si.html.

Fuller, Robert. 'Wonder and Spirituality', *The Journal of Religion*, vol. 86, no. 3, 2006, pp. 364–84.

Galliardetz, Richard R. (ed.). *When the Magisterium Intervenes: The Magisterium and Theologians in Today's Church*, Liturgical Press, Collegeville, MN, 2012.

Gilding, Paul. *The Great Disruption*, Bloomsbury Press, London, 2011.

Gillett, Carl. 'The Hidden Battles Over Emergence', in Philip Clayton (ed.), *The Oxford Handbook of Religion and Science*, Oxford University Press, New York, 2006.

Goodenough, Ursula. *The Sacred Depths of Nature*, Oxford University Press, New York, 1998.

Gottlieb, Roger. *Spirituality: What it is and Why it Matters*, Oxford University Press, New York, 2012.

Gregersen, Niels. 'Emergence and Complexity', in Philip Clayton (ed.), *The Oxford Handbook of Religion and Science, Oxford University Press*, New York, 2006, pp. 767–83.

———, 'Emergence: What is at stake for Religious Reflection', in Philip

Clayton (ed.), *The Re-Emergence of Emergence: The Emergentist Hypothesis from Science to Religion*, Oxford University Press, New York, 2006, http://site.ebrary.com/id/10271438?ppg=294.

————, 'The Cross of Christ in an Evolutionary World', *Dialog: A Journal of Theology*, vol. 40, 2001, pp. 192–207.

Grim, John, & Tucker, Mary Evelyn. *Ecology and Religion*, Island Press, Washington, DC, 2014.

Habel, Norman. *Rainbow of Mysteries: Meeting the Sacred in Nature*, Copper House, Kelowna, BC, 2012.

Hamilton, Clive. *Earth Masters*, Allen and Unwin, Sydney, 2013.

Harding, Stephan. *Animate Earth*, Chelsea Green Publishing, White River Junction, VT, 2006.

Haught, John F. *Is Nature Enough?*, Cambridge University Press, Cambridge, UK, 2006.

Hawking, Stephen. *A Brief History of Time*, Bantam Books, New York, 1988.

Hedlund-de Witt, Annick. 'Pathways to Environmental Responsibility: A Qualitative Exploration of the Spiritual Dimension of Nature Experience', *Journal for the Study of Religion, Nature and Culture*, vol. 7, no. 2, 2013, pp. 154–86.

Hines, Michael. 'Finding God in all Things: A Sacramental Worldview and Its Effects', in Landy Thomas (ed.), *As Leaven in the World*, Sheed and Ward, Franklin, TN, 2011, pp. 91–103.

Holman, Peggy. *Engaging Emergence*, Berrett-Koehler, San Francisco, CA, 2011.

Intergovernmental Panel on Climate Change (IPCC), 'Climate Change 2014: Synthesis Report', https://www.ipcc.ch/pdf/assessment-report/ar5/syr/SYR_AR5_FINAL_full_wcover.pdf.

John Paul II. *Dominum et Vivificantem: Giver of Life: On the Holy Spirit in the Life of the Church and the World*, 1986, http://w2.vatican.va/content/john-paul-ii/en/encyclicals/documents/hf_jp-ii_enc_18051986_dominum-et-vivificantem.html.

Johnson, Elizabeth A. *Ask the Beasts: Darwin and the God of Love*,

Bloomsbury Publishing, London, 2014.

———, 'From Pyramid to Circle, The Community of Creation', in Book Discussion, *Theological Studies,* vol. 77, no. 2, 2016, pp. 479–87, https://journals.sagepub.com/doi/10.1177/0040563916635120

———, 'Letter to Cardinal Wuerl, 14 July, 2011', in Richard R. Gaillardetz (ed.), *When the Magisterium Intervenes: The Magisterium and Theologians in Today's Church*, Liturgical Press, Collegeville, MN, 2012, pp. 248–52.

———, *She Who Is: The Mystery of God in Feminist Theological Discourse*, Crossroad Publishing, New York, 1992, 2002.

———, *Quest for the Living God: Mapping Frontiers in the Theology of God*, Continuum, New York, 2007.

———, 'To Speak Rightly of the Living God', in Richard R. Gaillardetz (ed.), *When the Magisterium Intervenes: The Magisterium and Theologians in Today's Church*, Liturgical Press, Collegeville, MN, 2012, pp. 213–51.

———, 'Turn to the Heavens and the Earth: Retrieval of the Cosmos in Theology', *CTSA Proceedings* vol. 51, 1996, pp. 1–14.

———, *Women, Earth and Creator Spirit*, Paulist Press, Mahwah, NJ, 1993.

Kasper, Walter, *The God of Jesus Christ*, Crossroads, New York, 1984.

Keller, Catherine. *On the Mystery: Discerning Divinity in Process*, Fortress Press, Minneapolis, MN, 2008.

———, 'No More Sea', in Dieter T. Hessel and Rosemary Radford Ruether (eds), *Christianity and Ecology: Seeking the Well-being of Earth and Humans*, Harvard University Press, Cambridge, MA, 2000, p. 193.

King, Ursula. 'One Planet, One Spirit', *Ecotheology*, vol. 10, no. 1, 2005, pp. 66–87.

Lane, Belden C. *Landscapes of the Sacred*, Paulist Press, New York, 1988.

Macy, Joanna. *Coming Back to Life*, New Society Publishers, Gabriola Island, BC, 1998.

McFague, Sallie. *The Body of God: An Ecological Theology*, Fortress Press, Minneapolis, MN, 1993.

———, 'An Ecological Christianity: Does Christianity Have It?', in Dieter T. Hessel & Rosemary Radford Reuther (eds), *Christianity and Ecology*, Harvard University Press, Cambridge, MA, 2000, p. 33.

McKibben, Bill. *Eaarth*, St Martin's Griffin, New York, 2011.

Mercy International Association, *Mercy International Reflection Process Guide Book: A Guide for using the Process and the themes of Laudato Si', On Care for Common Home.* Adele Howard (ed.), 2018. https://www.mercyworld.org/about/our-initiatives/mercy-international-reflection-process/mirp-handbook/.

Miller-McLemore, Bonnie J. 'Five Misunderstandings about Practical Theology', *International Journal of Practical Theology*, vol. 16, no. 1, 2012, pp. 5–26, https://doi.org/10.1515/ijpt-2012-0002.

———, 'Introduction: The Contributions of Practical Theology', in Bonnie J Miller-McLemore (ed.), *Wiley-Blackwell Companion to Practical Theology*, Wiley-Blackwell, Hoboken, NJ, 2012, pp. 1–20, http://site.ebrary.com?id?10500971.

Mooney, Christopher. 1991. 'Theology and Science: a New Commitment to Dialogue', *Theological Studies*, vol. 52.

Morwood, Michael. *It's Time: Challenges to the Doctrine of the Faith*, Kelmor Publishers, Sunbury, Victoria, 2013.

Murphy, Nancey. 'Introduction', in William R. Stoeger SJ and Nancey Murphy (eds), *Evolution and Emergence: Systems, Organisms, Persons*, Oxford University Press, 2007, pp. 1–16, http://site.ebrary.com/id/10271643?ppg=16.

Ormerod, Cynthia Crysdale & Ormerod, Neil. *Creator God, Evolving World*, Fortress Press, Minneapolis, MN, 2013.

Panikkar, Raimon. *The Cosmotheandric Experience: Emerging Religious Consciousness*, Orbis Books, Maryknoll, NY, 1994.

Paul II. 'Truth Cannot Contradict Truth', in Robert Russell, William Stoeger and Franscisco Ayala (eds), *Evolutionary and Molecular*

Biology, Vatican Observatory and Berkeley Centre for Theology and the Natural Science, Vatican City, 1998, pp. 2–9.

Peacocke, Arthur. 'Emergence, Mind and Divine Action: The Hierarchy of the Sciences in Relation to the Human Mind-Brain-Body', in Philip Clayton (ed.), *The Re-Emergence of Emergence: The Emergent Hypothesis from Science to Religion*, Oxford University Press, Oxford, 2006, pp. 272–94. .

———, *Theology for a Scientific Age: Being and Becoming-Natural, Divine and Human*, Fortress Press, Minneapolis, MN, 1993.

Phipps, Carter. *Evolutionaries: Unlocking the Spiritual and Cultural Potential of Science's Greatest Idea*, Harper Collins, New York, 2012.

Proctor, James D. *Envisioning Nature, Science and Religion*, Templeton Press, West Conshohocken, PA, 2009.

Rahner, Karl. 'Christology within an Evolutionary View of the World', in *Theological Investigations*, vol. 5, 1975, Seabury Press, New York, pp. 157–92.

Rasmussen, Larry. *Earth Community, Earth Ethics*. Orbis Books, Maryknoll, NY, 1996.

———, *Earth-Honoring Faith: Religious Ethics in a New Key*, Oxford University Press, Oxford, 2013.

Santmire, Paul. *Before Nature: A Christian Spirituality*, Fortress Press, Minneapolis, MN, 2014.

Schneiders, Sandra M. 'Approaches to the Study of Christian Spirituality', in Arthur Holder (ed.), *Blackwell Companion to Christian Spirituality*, Blackwell Publishing, Boston, MA, 2005, pp. 1–14.

Silberstein, Michael. 'Emergence, Theology and the Manifest Image', in Philip Clayton (ed.), *The Oxford Handbook of Religion and Science*, Oxford University Press, New York, 2006.

Steffen, Will et al. 'The Anthropocene: From Global Change to Planetary Stewardship', *AMBIO* vol. 40, 2011, pp. 739-61.

———, 'Planetary Boundaries: Guiding Human Development on a

Changing Planet', *Science*, January 2015, https://doi.org/10.1126/science.1259855.

Stoeger, William R. 'Reductionism and Emergence: Implications for the Interaction of Theology with the Natural Sciences', in Nancey Murphy and William R. Stoeger (eds), *Evolution and Emergence: Systems, Organisms, Persons*, Oxford University Press, Oxford, 2007, p. 244.

Swimme, Brian Thomas & Tucker, Mary Evelyn. *Journey of the Universe*, Yale University Press, New Haven, 2011.

Swimme, Brian & Berry Thomas. *The Universe Story: From the Primordial Flaring Forth to the Ecozoic Era – A Celebration of the Unfolding of the Cosmos,* Harper, San Francisco, 1992.

Tacey, David. 'Spirit Place', in John Cameron (ed.), *Changing Places: Reimagining Australia,* Longueville, Double Bay, NSW, 2003, p. 245.

Taylor, Bron. *Dark Green Religion: Nature Spirituality and the Planetary Future*, University of California, Berkeley, CA, 2010.

———, 'Religion and Environmentalism in America and Beyond', in Roger S. Gottlieb (ed.), *Religion and Ecology*, Oxford University Press, Oxford, 2006, pp. 588–612.

Teilhard de Chardin, Pierre. *Hymn of the Universe*, Harper and Row, New York, 1961.

Tilley, Terrence W. 'Book Review: Ask the Beasts: Darwin and the God of Love. By Elizabeth A. Johnson', *Theological Studies*, vol. 76, March 2015, pp. 194–5.

Tinney, Mary. *Ecology and Christian Faith*, Earth Link, 2009.

———, *When Heaven and Earth Embrace: How Do We Engage Spiritually in an Emerging Universe?*, PhD thesis, Australian Catholic University, 2018, https://researchbank.acu.edu.au/theses/716/

Treston, Kevin. *Emergence for Life not Fall from Grace: Making Sense of the Jesus Story in the Light of Evolution,* Mosaic Press, Preston, Victoria, 2013.

Trigger, Bruce G. *Understanding Early Civilisations*, Cambridge University Press, Cambridge, UK, 2003.

Turkson, Peter. 'Integral Ecology and the Horizon of Hope: Concern for the Poor and for Creation in the Ministry of Pope Francis', Trocaire Lenten Lecture, Maynooth, Dublin, 5 March, 2015.

Ungunmerr-Baumann, Miriam-Rose. 'Dadirri, The Spring Within', in Eileen Farrelly, *Dadirri: The Spring Within*: *The Spiritual Art of the Aboriginal People from Australia's Daly River Region*, Terry Knight and Associates, Nightcliff, NT, 2003, p. ix.

Veling, Terry A. *Practical Theology: 'On Earth as it is in Heaven'*, Orbis Books, Maryknoll, NY, 2005.

Wayman, Erin, 'What's in a Name?', Smithsonian.com, 16 November 2011, https://www.smithsonianmag.com/science-nature/whats-in-a-name-hominid-versus-hominin-216054/#mAI3lZOzILQHW8kZ.

White, Lynn. 'The Historical Roots of our Ecological Crisis', *Science*, vol. 155, March 1967, pp. 1203–07.

Wolfteich, Claire. 'Animating Questions: Spirituality and Practical Theology', *International Journal of Practical Theology*, vol. 13, no. 1, pp. 121–43, https://doi.org/10.1515/IJPT.2009.7.

SUPPLEMENT

Guides for Reflection and Conversation

Principle 1

Principle 2

Principle 3

Principle 4

Principle 5

Principle 1: Vision and Mission

Recognise that the universe is a dynamic entity within which all is interconnected yet distinct.

We are in a period of disequilibrium, and the systems of thought that gave us meaning are breaking down. Is this your experience?

How does Earth Link's vision of a world in which there is respect, reverence and care for the whole community of creation speak to this situation?

To what extent do you identify with Earth Link's mission to 'facilitate deep bonding within the whole Earth community' as a way of living into our current situation?

In the past week, how conscious have you been of the relationship between cosmos, Earth and humans? How do you envision that relationship?

Principle 2: Mystical Encounters

*Cultivate an open, attentive and receptive attitude
in order to enter into transformative, mystical encounters.*

Ecospirituality is

- cosmic in scope, connecting Spirit with all of life
- attentive to the sacred inner dimensions of nature
- open to transformative, mystical encounters with nature
- motivated for justice for the whole Earth community.

Which of these dimensions is strongest in your ecospirituality?

When did you last experience wonder or awe? How easily do you subscribe to Martin Luther's insight that the finite bears the infinite?

Nature provides guidance about how to think of the presence and activity of the Spirit of God in the natural world, a Spirit that loves, pervades and vivifies, while remaining transcendent, incomprehensible Mystery. How can nature provide this guidance?

Do you have a regular spiritual practice of immersion in nature?

Principle 3: Indwelling Presence

Acknowledge the indwelling presence
of the Spirit in Earth and cosmos.

In the poem *God's Grandeur*, Gerard Manley Hopkins says, 'The world is charged with the glory of God.' Have you had this same conviction? In what situation?

The indwelling Spirit does not intervene in our created reality but accompanies the community of life at every moment. This a change in the way we view God's presence. What difference does this make to the way you pray?

Theologically there is a difference between panentheism, understood as God being present and engaged with all that is, although God is over and above all, and pantheism, according to which God is identical with created reality. The Christian position favours the former because it respects both God's transcendence and God's immanence. How do these understandings affect ecospirituality?

It is a basic evolutionary principle that reality is unfolding or emerging according to principles within it and does not require ongoing direct intervention by God in order to evolve. How is this scientific understanding compatible with our experience of God as a God of love who empowers rather than exercises power over?

Principle 4: Jesus In Solidarity

Acknowledge that in Jesus Christ the God of love embraces the cosmos in healing solidarity.

How do you make meaning of the evolutionary realities of pain, suffering and death?

In assuming flesh, Jesus Christ assumes the cosmos. Jesus enters into solidarity not only with humans but with the whole created reality. He embraces its biological processes and connects to the stuff of the universe. How does this realisation affect the way in which we understand the Incarnation of Jesus and his redeeming mission?

In Jesus there is divine solidarity with suffering and death, pointing to God's liberating, healing and inclusive love. God accompanies us. This embrace does not justify suffering as an end in itself but recognises it as sometimes inevitable and often to be redressed. How does this understanding differ from seeing the life, death and resurrection of Jesus as atonement for our sins?

Creation theology reminds us that matter does not explain or ground itself but is a gift from the Giver of life. God's loving embrace is also luring us towards fullness of life. What light does this cast on the experience of death?

Principle 5: The Basis of Right relations

Live in right relations within the community of creation.

We need to move beyond a paradigm of dominion that stresses human superiority to one of communion that locates humans within the community of life, interconnected with all that is. What shifts in your thinking might this entail?

We are part of an interwoven community of creation praising God and giving glory by being our distinctive selves. What responsibilities flow from this?

We are called to ecological conversion, which Elizabeth Johnson describes as falling in love with earth as an inherently valuable, living community in which we participate, and bending every effort to be creatively faithful to its well-being, in tune with the living God who brought it into being and cherishes it with unconditional love' (*Ask the Beasts,* p. 259). In this quote, what motives does Johnson identify for such a conversion?

What are you doing to facilitate the embrace of Heaven and Earth and address the negative effects of human influence on Earth and cosmos?

Index

Milton Keynes UK
Ingram Content Group UK Ltd.
UKHW021939281024
450365UK00018B/1166